The Thing About Leftovers

C. C. Payne

SCHOLASTIC INC.

This is a work of fiction. Names, characters, places, and incidents either are the product of the author's imagination or are used fictitiously, and any resemblance to actual persons, living or dead, businesses, companies, events, or locales is entirely coincidental.

No part of this publication may be reproduced, stored in a retrieval system, or transmitted in any form or by any means, electronic, mechanical, photocopying, recording, or otherwise, without written permission of the publisher. For information regarding permission, write to Puffin Books, an imprint of Penguin Young Readers Group, a division of Penguin Random House LLC, 375 Hudson Street, New York, NY 10014.

ISBN 978-1-338-24961-3

Copyright © 2016 by C.C. Payne. All rights reserved. Published by Scholastic Inc., 557 Broadway, New York, NY 10012, by arrangement with Puffin Books, an imprint of Penguin Young Readers Group, a division of Penguin Random House LLC. SCHOLASTIC and associated logos are trademarks and/or registered trademarks of Scholastic Inc.

The publisher does not have any control over and does not assume any responsibility for author or third-party websites or their content.

12 11 10 9 8 7 6 5 4 3   19 20 21 22

Printed in the U.S.A.   40

First Scholastic printing, October 2017

Design by Annie Ericsson

For Mark Payne, who—miraculously—chose me,
as if I were the best thing on the menu:
Thank you.
All my love forever,
C.

Chapter 1

There's nothing worse than leftovers, except school-cafeteria leftovers, which are so bad they should be called something else—leftunders maybe? Yesterday the school had served tacos for lunch and today it was sloppy joes. The sloppy part tasted like leftover taco meat mixed with tomato sauce and slapped on a soggy white bun, which is exactly what it was. Yuck.

As usual, I had to face my leftunders alone. Okay, I wasn't *technically* alone. There were other girls at my lunch table, and they were all very . . . polite. But they never spoke to me unless I spoke to them first. They didn't seem to dislike me exactly; they just didn't seem to need any new friends—*nobody* in Lush Valley seemed to need any new friends. We were more than halfway through the school year now and I'd long since given up the hope that I'd enjoy lots of friends, birthday parties, and sleepovers in Lush Valley—like I had in my old neighborhood.

I'd told myself that was okay, that my old friends were enough for me, too, and I'd tried to hang on to them. My last attempt to hang on was trying to throw a slumber party at my old house, which was Dad's house by then. But when I called my friends, one by one, to invite them, nobody could come. None of them said why. Finally my best friend, Olivia Moore, told me that everyone was already going to a slumber party at Liddy Duncan's house on that night. "Do you want me to ask Liddy if you can come?" Olivia said, sounding uncomfortable. "Um . . . no, thanks," I'd said, because I barely knew Liddy and it was all too awkward—I have enough awkward in my life already.

I squeezed my eyes shut against the memory and took a deep breath.

When I opened my eyes, I found myself looking at the popular girls' table: They were all eating various kinds of sushi from bento boxes, which are especially made for packing delicate sushi and sauces. Everything about the popular girls, from their designer shirts and jeans to their lunches, said *Lush Valley*.

Just then, the leader of the popular girls, Buffy Lawson, caught me staring.

I returned my attention to my tray, wondering what my lunch said. I decided my lunch also said the same thing as my clothes: *leftovers*. But hey, at least I had the courage to look like myself. Not that I wouldn't have appreciated a few more options for looking like me, but at least I didn't look like I'd come straight out of the Lush Valley cookie cutter for kids.

When lunch was almost over, Buffy gave her friends a little nod, then stood and made her way to the wide aisle in the center of the cafeteria. Naturally, all of her friends followed, for what I think of as their "big catwalk moment." One by one, they all strut up the center aisle, behinds swishing, to the trash can at the front of the cafeteria, where they strike a subtle pose as they dump their trash, then pivot and walk back, like fashion models on a runway—which they probably think they are—especially Buffy, who actually says, "I just *live* for high fashion!" (I'm pretty sure that anything with ruffles qualifies as "high fashion" by Buffy's standards because she wears a lot of ruffles. *A lot.*)

I don't do ruffles, bows, flowers, or sequins—I am against the cruel treatment of clothes by way of bedazzling. I have flannel shirts for winter, T-shirts for summer, and long-sleeved T-shirts and baseball-type shirts for everything in between—I don't devote a lot of thought to clothes because I like to think I have more important things to think about. Even so, I wouldn't mind some designer jeans; only Mom won't buy them, which probably explains a lot about my situation here in the valley—I don't have the right stuff, literally or figuratively.

I stood and picked up my lunch tray, reminding myself what my aunt Liz had told me: "Almost everybody who's anybody was nobody back when they were in school."

Okay, so maybe I'm a nobody at school, but the way I look at it, that just means I'm on my way to becoming somebody. And I am. I'm on my way to becoming a brilliant, world-famous chef who will have her own cooking show on television. That is, after I have my freckles surgically removed. Yes, by then I'm sure doctors will have made great breakthroughs in the area of freckles.

My TV show will be called *Fabulous Foods and Feasts with Fizzy Russo*—that's me. (My real name is Elizabeth, but everybody's been calling me "Fizzy" since the age of two, on account of that's what I called myself, and it seemed to fit my bubbly personality—back then, at least.) Anyway, on my TV show, I'll have a hundred matching clear-glass bowls in every size, in which I'll put beautiful, colorful ingredients, placing them in perfect order in a semicircle before me—this is called *mise en place*, which is a fancy French cooking term meaning "everything in its place." I'll also have great big jeweled rings on my fingers for when the cameras zoom in on my hands while I work—jewelry, not clothing, is meant to be dazzling! Shoot, after I appear on TV, Buffy Lawson will probably want my autograph, but I won't give it to her. I won't. I really won't.

That's what I was thinking as we all lined up in the cafeteria. I got behind Christine Cash, one of Buffy's followers, and met her eyes, thinking, *And you'll want my autograph, too, Christine.*

As if she could read my mind, Christine rolled her eyes and turned her back.

Okay, so Christine didn't want my autograph today. That didn't mean she wouldn't want it *ever*.

Back in the classroom, our French teacher, Miss Fehr, finished her lesson on telling time. Apparently, the French complicate time-telling with math. For example, if it's 5:40, they don't say, "It's five forty." They say, "It's five forty-five minus five"—in French—so you have to translate the language *and* do your own math! Since I hate math, I made a mental note to be sure and take my watch with me to Paris—when I go for culinary school—and spent the rest of class staring out the window.

It was a perfectly sunny January Thursday in Louisville,

Kentucky—except for the bitter cold—and I knew what that meant: It meant that Coach Bryant would take us outside for gym class. Coach Bryant is big on fresh air and exercise. According to him, fresh air and exercise would solve most of the world's problems.

It was then that I realized I'd accidentally worn gym shoes. See, if you don't wear gym shoes on gym days, then you don't get to participate in gym class. So, occasionally, I forgot to wear gym shoes—on purpose. On those days, Coach Bryant sent me to the library to write a one-page report on whatever we were studying—football (first played in England), soccer (similar games were documented as early as 50 B.C.), basketball (invented by a Presbyterian minister), or whatever. I usually finished my report in about twenty-five minutes—hey, I didn't say it was a *great* report—and then I had another twenty-five minutes to myself to read cookbooks.

Here's the thing: I'm not exactly athletically inclined. In fact, I'm pretty sure there's a magnetic field in my nose that attracts all athletic balls within a half-mile radius directly to it. But I'd just have to risk my nose today, since I had on gym shoes and all.

Coach Bryant was waiting by the double doors inside the gym, kickball in hand. As soon as the bell rang, he blew his whistle, shouted, "Everybody line up," and pushed out into the bright-white sunlight.

Of course, I was the last kid picked for a kickball team. As I stood on the blacktop with my hands in my coat pockets, waiting for my turn to kick—and embarrass myself—a sick feeling seeped into my stomach.

I tried to distract myself with the thought of dinner: What would I cook tonight?

I've been in charge of dinner—mostly—ever since my mom and dad divorced. Lots of things changed A.D. (after divorce), including my former stay-at-home mom, who now works full-time selling advertising for the *Courier-Journal*, which is the newspaper here in Louisville. Pretty much ever since she started working, Mom's been "running late," as she says. She's late for work, she's late getting home, and she used to be *very* late starting dinner.

After a few weeks of eating canned soups and grilled cheese sandwiches for dinner, I'd volunteered to cook. Now, that may sound like big stuff, but here's a little secret: Anyone who can read and follow directions can cook. Of course, I'd had to prove that I wasn't going to start a fire using the stove, that I'd be careful with the knives, and so on, but after that, the job was all mine.

I love my job. I love reading cookbooks, planning our meals, and making the grocery list—honestly, I love making a list of any kind. I love grocery shopping with my mom every other Saturday morning. And the cooking . . . well, cooking dinner is my favorite part of any day. Plus, I only cook things that *I* like!

I was thinking of making us Kentucky Hot Browns—country ham and turkey on top of Texas toast, covered in a heavy cream sauce, topped with bacon and tomato, and smothered in melted cheese—because Aunt Liz had recently given me the original recipe from the Brown Hotel. Did we have any heavy whipping cream at home? I tried to think, running through the contents of our refrigerator in my mind.

My heart sank, because that's when I remembered we still had enough red wine vinegar chicken leftovers from last night to make another meal. That meant I wouldn't get to cook today. That was the rule: Mom said if there were enough leftovers to make another meal, then we had to eat them—because food is money, and we can't afford to throw away money.

It's hard to make a meal for only two people and not have any leftovers, especially since most recipes are written for at least four people—usually more. But that's where my new cookbook came in. Yesterday, Aunt Liz had lent me a cookbook called *Two for Dinner* and all the recipes were written for just two people! I was hoping the book would put an end to leftovers at my house once and for all. I'd planned to start reading it right away—today—during gym!

"Hey, you're up!" someone yelled.

I blinked. It was my turn to kick.

Mike Anderson, the boy pitching, made a show of rolling the ball to me in super slow motion.

I ignored him, took a few running steps, and kicked as hard as I could. I was thrilled when the ball connected with my foot and went flying through the air. I took off running for first base as fast as I could.

The class groaned.

When I arrived at first base, Jimmy Cox, the boy standing on it, turned to me and said flatly, "Foul."

I swiped at my eyes and sniffed. The cold, or maybe the running, or both, had caused my eyes to water and my nose to run. I willed my face not to leak as I walked back to home base, to take my place at the back of the line.

"You have to kick again," Christine Cash said, rolling her eyes.

Why? Why couldn't I just be done?

Again, Mike made a show of rolling the ball to me as if I'd just learned to walk, and again, I took a running start and kicked.

This time, the ball flew directly into the hands of an outfielder and all the outfielders moved in to change positions with us. It was our team's turn in the outfield. Now, the only thing I hate worse than kicking is fielding.

My team completely ignored me while I tried to decide where the ball was least likely to go so I could stand there. But sure enough, on the first kick, before I even knew what was happening, the ball smacked me hard in the face and blood poured out of my nose—worst facial leakage ever.

"Gross," someone said.

I covered my nose with both hands and glanced over at our team captain, Buffy Lawson—who else?

Buffy held up her hands and said, "Fizzy, these are called hands. Maybe you could try using yours next time."

I removed one—slightly—bloody hand from my face, reached out like I was going to touch her, and said, "Like this?"

Buffy squealed and skittered away from me as the wind blew a big chunk of her perfectly styled hair into her perfectly sticky lip gloss. She swatted at it and glared at me like I was responsible for this tragedy.

"Okay, okay," Coach Bryant said to Buffy. Then he looked over at me and shook his head like he couldn't believe this had happened—again.

At least I was out for the rest of the game. I leaned against the brick wall, holding an ice pack to my nose and promising myself two things: 1) that I'd remember to "forget" my gym shoes next week; and 2) that I'd stop by Aunt Liz's house on my way home from school.

Chapter 2

Whenever I have a bad day at school, I stop at my aunt Liz's house afterward. I walk right by Chrysanthemum Court—where Aunt Liz lives—on my way home anyway. But even if I didn't, I'd walk far out of my way to get there. Somehow Aunt Liz always figures out what's squashing me, and she's usually able to lift it, set me straight, and fluff me back up again—like one of her decorator pillows.

That afternoon, Aunt Liz was working at the desk in her front window. As soon as she spotted me through the window, she jumped up and hurried out to meet me. She was much fizzier than I, I noticed. And prettier, too: Aunt Liz is tall and thin, with long, heavy dark hair, dark skin, dark eyes and eyelashes.

I was named after Aunt Liz on account of her brother is my dad. He is also dark and handsome. But he married a light-haired, light-skinned lady with green eyes: my mother. The result was, of course, me.

I am pretty weird looking: My body is way too skinny for my normal-size head, which is covered with strawberry-blond hair that is more strawberry than blond. I have lots of freckles, especially on and around my nose, which has a permanent bump on the bridge—thanks to soccer balls, kickballs, volleyballs, basketballs—you get the picture. My eyes are green. I think they're okay, but oh, how I wish I had Aunt Liz's dark hair and skin! Sometimes I wish I *were* Aunt Liz.

"I was hoping I'd see you today," Aunt Liz said, giving me a little squeeze.

"How come?" I asked.

"My copy of *Southern Living* just arrived," Aunt Liz said, holding up the magazine, "and in it is everything you need to know about the *Southern Living* Cook-Off this year!"

It had been my dream to win the *Southern Living* Cook-Off ever since I'd first laid eyes on the magazine. Cook-off winners were announced in that very first issue I saw, and their winning recipes were in there, too—all delicious! That same issue had also featured Aunt Liz's new sunroom and the garden beyond. (Aunt Liz is an interior designer; she decorates most all the fanciest houses in Lush Valley.) Yep, my very own aunt Liz is practically famous thanks to *Southern Living*. I'd loved the magazine ever since.

"I'm Southern, right?" I asked.

Aunt Liz laughed. "Yes, honey, you are as Southern as grits."

"So I'm a Southern Italian?"

"No," Aunt Liz said. "You're a Southern American with some Italian heritage."

I clapped my gloved hands together. "My cheese grits! I should send *Southern Living* the recipe for my cheese grits!"

Aunt Liz nodded and said, "Well, we can't make important decisions like this on an empty stomach. C'mon in and we'll find you something to eat—it's freezing out here!"

I followed her into the house, admiring the way her long, drapey sweater floated around her. I wished I had a long, drapey sweater like Aunt Liz's.

I also wished I had a kitchen like hers. For me, there is no better place on earth than Aunt Liz's kitchen. It's big and roomy, but somehow still feels warm and cozy. There are lots of windows and plants and cookbooks; the walls are the color of butter, and it always smells sweet, like there's a cake in the oven. I'm considering filming my TV show there.

As usual, while Aunt Liz went to work fixing a snack for me, I picked up the kitchen phone and dialed my mom's office.

"Cecily Russo," my mother chirped.

"Hi, Mom," I said. "I'm at Aunt Liz's house."

"All right. How was your day?"

"Fine," I said, because this is what I say every day, no matter what.

"Just a sec," I heard my mother say in a muffled voice that let me know she had covered the phone with her hand and wasn't talking to me.

I waited for a few seconds and then asked, "Red wine vinegar chicken again tonight?" I was hoping—hard—that she'd say no.

"No, I don't think so. . . . Yes, I know—I'll just be a sec," Mom said.

"Are you talking to me?" I asked.

"Yes, you: no chicken," Mom said as I heard the door to her office click closed in the background.

"So I get to cook?" I perked right up at that possibility.

"Uh, no... I've invited Keene to join us for dinner, so I'll be cooking," Mom said.

Keene Adams is my mom's boyfriend and I hate it when he comes to dinner. For starters, it means that I don't get to cook, and that's just for starters.

"Oh," I said quietly. "Well... maybe I could just stay here." I looked over at Aunt Liz, wearing a question on my face.

Aunt Liz nodded.

"Did she invite you to stay?" Mom asked, because she's always worried about me inviting myself places, which I never do, except with Aunt Liz and she doesn't mind.

"Yes," I said.

"Yes, what?" Mom said.

"Yes, ma'am," I said, because Mom is big on *ma'ams* and *sirs* and *pleases* and *thank-yous*, and manners in general.

After everything was settled and I'd put the phone back, I shucked my coat and sat down at Aunt Liz's kitchen table. Aunt Liz placed a tall glass of sweet tea and a Benedictine sandwich in front of me—Benedictine is a thick spread made from cream cheese, cucumbers, onions, and mayonnaise that was created right here in Louisville, at Benedict's Tea Room.

I smiled. The fact is I could probably live on sweet tea alone, but I could *surely* live on sweet tea and Benedictine because... well, sweet tea is *sweet tea*, and Benedictine tastes so light and clean and fresh, like spring in your mouth! Yum!

"Thank you," I said.

Aunt Liz nodded as she sat down across from me and seemed to wait for me to say more.

"Keene is coming to dinner," I announced, crinkling my nose.

"I kind of figured that," Aunt Liz said, smiling. Then her face went all serious as she leaned over the table and asked, "Why don't you like him, Fizzy?"

I shrugged.

"I bet he likes you," she offered.

"He doesn't," I informed her.

Aunt Liz's dark eyebrows knitted together above her pretty nose. "What makes you think that?"

No way was I going to tell her. No way. What if I told her and then she agreed with Keene? What if Aunt Liz stopped liking me?

Aunt Liz's face relaxed as she leaned back in her chair. "You don't have to tell me if you don't want to."

I took a bite of my sandwich.

"Okay," Aunt Liz said, picking up the magazine again and flipping through the glossy pages. "Recipes must be mailed by February fifteenth. Each recipe will be tried in the *Southern Living* test kitchen and finalists will be notified by mail no later than June first."

I swallowed and said, "I want to get my recipes in as soon as possible . . . so they can try them and get back to me—quick."

"That doesn't mean they will," Aunt Liz said. "Even if they like them, they'll have to compare them against thousands of other entries."

I took another bite of my sandwich.

"There are five categories in the cook-off," Aunt Liz read aloud from the magazine.

I chewed and nodded for her to go ahead.

"The categories are Your Best Recipe, Family Favorites, Southern Desserts, Healthy and Good for You, and Party Starters."

"I'm entering all of them," I informed her.

Aunt Liz smiled and nodded like she'd known it all along. Happy, excited feelings filled the kitchen, and I wished I could stay . . . forever.

Chapter 3

I call my *Sports Illustrated* alarm clock "Genghis," as in Genghis Khan, dreaded emperor of the Mongol Empire, who once said, "I am the punishment of God . . . If you had not committed great sins, God would not have sent a punishment like me upon you." Believe me, Genghis is a terrible punishment who came disguised as a birthday present from my mom's boyfriend, Keene. Genghis might as well punch me in the stomach to wake me up every morning, because I always feel sick when he starts screaming. That Friday morning was no different. But then, I feel sick a lot.

The sickness started when Mom and I moved to Lush Valley last May—right after school let out, and after Mom and Dad's divorce was final. Since Dad got the house in the divorce, we had to move. But the sickness didn't start right away. No, at first, I'd been happy. Being in a new place seemed exciting, adventurous, almost like a vacation. But after a while, the excitement wore off. It was like that time at the fair when I'd had my fill of funnel cake and cotton candy and I'd ridden the Tilt-A-Whirl one too many times: Suddenly all the hot, sweet scents made me feel pukey and I just wanted to go home. Then and now. I missed my dad, my old house, my old neighborhood, my old friends, my old school—everything. That's when the sickness came.

And even though my parents had lived apart for more than a year by then, and even though they didn't seem to like each other—at all—I still told myself that they'd eventually get back together. And then I'd get to go home. I even believed it. I believed it right up until my dad got remarried last August—and I think Mom might've believed it, too, because I saw tears well up in her eyes when I told her that Dad was getting married.

But now I understand that my parents aren't ever going to get back together. I blame my dad's new wife, Suzanne, for this. Oh, sure, I know Dad's at fault, too, and at first, I blamed him *and* Suzanne and was mad at them both. But it's hard to stay mad at someone you love the way I love my dad. Suzanne, on the other hand . . . well, I've been able to stay mad at her just fine. (The family counselor we saw—a few times—explained all of this and said it was completely normal for me to blame Suzanne, even though none of it was Suzanne's fault or Dad's fault or Mom's fault. He said that *no one* was at fault. Over and over again. But I figure it *has* to be somebody's fault—because I'm pretty upset.)

Anyway, this morning when Genghis started screaming at me, I sat up in bed, reached over, and gave him a good, hard WHAP! I eyed the clothes hanging on my closet door. I knew they weren't the "right" clothes, but even so, they were the right clothes for me. Well, except for the same plain jeans that Mom's been buying on sale at the end of each season for my entire life. I've tried to show Mom the error of her ways, but she just says stuff like "Things aren't important and what other people think about our things certainly isn't important. *People* are important, and we only moved to Lush Valley because it's the best school district!" I'm pretty sure nobody else around here thinks that way—Lush Valley is often called "Luxe Valley" due to all the large, luxurious homes here. We don't live in any of those. We live in a small, two-bedroom town house, with one full bathroom upstairs—which I share with Mom—and a half bathroom downstairs.

I dragged myself out of bed and found Mom blow-drying her hair in the bathroom.

When she saw me, she turned the dryer off and said, "I'm running late."

My stomach did a Tilt-A-Whirl. "What happened?" I asked.

"What do you mean?" Mom said.

"Why are we running late? Why are we *always* running late?"

Mom stiffened. "I'm doing my best, Fizzy. I'm trying to be the best mother, housekeeper, saleswoman, girlfriend, and friend I know how to be. It takes extra effort—and time—to be the best at one thing, let alone five things. It's hard. And I don't appreciate being questioned by my own daughter."

"Yes, ma'am. I'm sorry," I said, and I felt relieved when Mom turned the blow-dryer back on.

I hurried through the empty hallway as fast as my legs would carry me even though I knew I was too late. Again.

I was right: When I reached my homeroom, the door was

locked. I could see Mr. Moss and the rest of my science class through the little slice of window in the door, but when I knocked, Mr. Moss merely glanced at me, shook his head at the boy sitting nearest the door, and went on teaching. That meant I had to go down to the principal's office. Again.

Mrs. Warsaw, the principal, seemed to be waiting for me. She wore a tight look of disapproval and a navy skirt suit with pearls. When she saw me, she uncrossed her bony arms, said in a clipped voice, "Come with me, please, Elizabeth," and took off walking.

I followed her into her office and sat down when she pointed at a chair.

Mrs. Warsaw harrumphed into the seat behind her spotless desk and opened a folder. "Elizabeth, do you realize this is your ninth tardy?"

"No, ma'am," I said.

Mrs. Warsaw looked up from the folder. "Well, *it is*," she said, like I didn't believe her.

I believed her. Really, I did. "I'm sorry," I said.

Mrs. Warsaw sighed. "Why are you late?"

I thought about this. "I . . . I don't know . . . but my mom says she's doing her best."

Mrs. Warsaw's eyes narrowed. "Are you trying to blame your mother for your tardiness? Perhaps I should call her."

"Please don't do that," I said quietly. "She's very busy."

Mrs. Warsaw looked satisfied as she slapped the folder shut.

I tilted my head back and blinked at the speckled rectangles on the ceiling. I was trying to use gravity to force back the tears in my eyes, because I don't cry in public. Sometimes I want to, but I don't do it.

"Did you just roll your eyes at me, young lady?"

"No, ma'am," I said quickly, jerking my chin down to stare at Mrs. Warsaw through shocked-wide eyes.

"Well. From now on, Elizabeth, I expect you to be on time. Is that understood?"

"Yes, ma'am."

"Get yourself an alarm clock. And use it." Mrs. Warsaw stood. I didn't bother telling her about Genghis, which was just as well, as she continued, "You'll receive a tardy slip on your way out."

I pressed my—unexcused—pink tardy slip against the window so that Mr. Moss could see it when I knocked on the door. Again.

An hour later, during English class, a note arrived with my name on it and it turned out that the guidance counselor, Mrs. Sloan, wanted to see me.

Mrs. Sloan looked up from her messy desk when I knocked—at least I *think* there was a desk under there somewhere—and smiled like she was downright thrilled to see me. "Come in, Fizzy," she said. "Come right in."

Mrs. Sloan was the opposite of Mrs. Warsaw: She had long, wild, curly gray hair and wore loose layers of clothes over her soft curves, with lots of big, bold jewelry—something about her made me think gypsy.

I did as I was told and shut the door behind me—so that no one would see me in the guidance counselor's office.

"Please sit down," Mrs. Sloan said, coming out from behind her desk and motioning toward a small round table with three chairs.

"No, thank you," I said. "I don't ... um ... need to be here."

Mrs. Sloan's eyebrows moved up but her smile stayed put. "And why is that?"

"I don't have ... you know ... *problems*."

"Everybody has problems," Mrs. Sloan said easily.

"I don't," I insisted.

Mrs. Sloan perched on the edge of her desk. "You don't think you have any problems? Do you think that's a problem?"

"Are you trying to *make* problems for me?" I asked. "Because I don't need that—I already ..." I stopped talking and dropped my eyes to the cheerful area rug.

"Already have enough problems?" Mrs. Sloan guessed. "That's okay. Everybody has problems. We can talk about it—maybe I can help."

Out of desperation, I blurted, "I've already seen a family counselor and he said that I'm a perfectly healthy, perfectly normal girl who just needs time to adjust."

"Great," Mrs. Sloan said. "And how do you think that's going?"

"Fine. I'm fine. It's all fine."

"Then why can't we talk about it?" Mrs. Sloan asked.

I gave her an exasperated sigh. "Because that wouldn't be polite ... or ladylike."

"So, a polite lady doesn't . . . ?" Mrs. Sloan held up her hands, shrugged, and waited for me to finish her sentence.

I huffed, "Doesn't finish other people's sentences, doesn't discuss unpleasant or private stuff, and doesn't get all . . . *emotional*—because that makes other people uncomfortable."

The smile slipped from Mrs. Sloan's face as she drew back, surprised. For a few seconds, she just stared at me like I was a new species she'd never before encountered. Then she nodded her understanding.

I turned to go.

"Fizzy?" Mrs. Sloan said as my hand closed around the doorknob.

I looked at her.

"This is one place you don't have to worry about being polite. You don't have to worry about anyone else's feelings or discomfort. You can say whatever you want without fear of judgment or consequences. And nothing you say will ever leave this room."

"I'm fine," I said. "Tell everybody." And then I went back to class.

For the rest of the day, I wondered what I might've said or done that had resulted in Mrs. Sloan sending for me. The tardiness was all I could come up with. So I knew I couldn't be late to school anymore.

Chapter 4

I barely cried at Aunt Liz's house that afternoon—we're talking one tear, maybe two, max, slipped from my eyes as I sat at her kitchen table spilling my guts—and very little else.

"The principal, Mrs. Warsaw (*sniff*), hates me," I told Aunt Liz. "I mean, she *really* hates me. She hates me so hard and so much that I'm pretty sure even if I jumped in front of bus or something to save her life, she'd still hate me."

Aunt Liz gave me a sympathetic little smile and said, "I'm sure she doesn't hate you, Fizzy."

"She sicced the guidance counselor on me. I think."

"*Oh*," Aunt Liz said. "What did the guidance counselor say?"

"Just that I could talk to her."

"Doesn't sound very attack dog–like."

I shrugged. "She was nice, I guess—nosy, but nice."

"So you talked to her?"

"Not exactly—mostly I just explained to her about manners."

"Manners?" Aunt Liz repeated.

"Yeah, you know, how you're not supposed to talk about private family stuff outside the family, or whine and complain about things, or get all emotional and scream and cry and snot all over yourself—because it's not polite."

Aunt Liz laughed.

"I don't think Mrs. Sloan—the guidance counselor—is from Lush Valley."

Aunt Liz laughed again, but her face was serious when she said, "Fizzy, sometimes it's very helpful to discuss family and feelings. Believe me, people do it every day."

"Yeah . . . in places like California. I watch TV, too."

Aunt Liz smiled and shook her head.

"And, anyway, I have you to talk to."

Aunt Liz reached across the table, covered my hand with hers, and said, "You do. Always. Please remember that, Fizzy. But if you ever want to talk things over with your guidance counselor, I think that's okay, too."

"Well, I don't. So I can't be late to school anymore."

Aunt Liz took her hand back and thought about this. "You know, you could walk to school in the mornings just like you walk home in the afternoons—it's the same distance no matter what the time of day."

Why hadn't I thought of that?

"I bet that would help your mom, too."

Right again.

For a minute I just sat there wondering why I hadn't walked both ways from the beginning. I guessed riding with Mom in the mornings was a habit that had followed us from our old house, our old life. Mom used to drop me off at my old school in the mornings—I rode the bus home with my friends—but she hadn't worked back then, and my school hadn't been anywhere close to home. Things had changed.

Then I thought about how walkers are much cooler than car-riders—nobody looks cool being dropped off at school by their parents. I mean, car-riders might as well arrive caged or leashed for all the choice they have in the matter. But walkers? They could leave the house and go anywhere. Walkers *decide* to come to school. Plus, nobody knows for sure if walkers even *have* parents—parents are very uncool—and if they do, well, nobody has to *see* them. Yeah, I definitely wanted to be a walker, I decided. *Tick, tick, tick,* whispered the old wall clock. I looked at it: four o'clock.

I shot up out of my chair. "Oh! I have to go home and pack! It's Dad's weekend!"

"Right—just let me get the recipes," Aunt Liz said, jumping up and rushing to the island, which was littered with cookbooks and cards and pens and highlighters.

"Recipes?"

"Yes, I've been looking through my recipes and I found a few I thought you might be interested in—for the contest."

The cook-off. I'd almost forgotten! "Thanks," I said, jamming my hands into my gloves.

"Now, what you should do is try these out and then if you like them, you can play with them," Aunt Liz said. "Try adding different ingredients—more of this, less of that—you know."

I nodded. I love playing with recipes.

Aunt Liz handed me three index cards with recipes written on them in her lovely, loopy handwriting.

I hugged her.

She kissed the top of my head and said, "Those are just desserts. I'm still looking for other things to fit the other categories."

"Thanks, Aunt Liz."

I always forget something when I pack for my dad's house. No matter how long I take doing it or how hard I try to remember everything, I always forget something. Lots of times, I've forgotten my toothbrush, but luckily Dad's a dentist, so he has plenty of those. Once, I forgot my tights. Last time, I'd forgotten my church shoes. Of course, Dad and Suzanne still made me go to church, wearing sneakers with my Sunday dress—humiliating! So I made myself a packing list, and number one on that list read *church shoes*.

I saw Dad's car pull to the curb through my bedroom window. I stuffed my list, along with Aunt Liz's recipe cards, into the front pocket of my suitcase and zipped all the zippers. Then I lugged my suitcase down the stairs, grabbed my coat, and locked the front door on my way out.

The homesick feeling stabbed at me as I walked past the bank of mailboxes, all of which were small and required a key. The mailbox at my old house—Dad's house—was huge, and I used to shove all kinds of things in it when I was too busy playing with my friends outside to make a trip inside: empty water bottles; books; sidewalk chalk; sweaters; shoes, shoes, and more shoes. (I'm not allowed to go outside without shoes on, but . . . well, my feet like to be free.) It used to be a running joke at our house that one just never knew what they might find in the mailbox: Lost your car keys? Check the mailbox. Out of

Scotch tape? Maybe some will turn up in the mailbox later. But my new mailbox only holds mail. And mailboxes aren't fun or funny to anybody anymore, least of all me.

"Did you remember your church shoes?" Dad asked as soon as I was in the car—he must've been as humiliated by my Sunday sneakers as I was.

Note to self: Church shoes are important to Dad, too, I thought. "Yes, sir," I said, buckling my seat belt and settling in for the half-hour ride.

Dad nodded his approval. "How was school?"

"Fine," I lied.

The car was pretty quiet after that. I looked out the window, watching skeletal trees whiz by and noticing how dull and gray winter really is, once Christmas is over.

Dad announced, "Suzanne and I have to attend a business dinner tonight—I'm thinking of taking on a partner—but we shouldn't be gone more than two hours."

I turned from the window. "Okay."

"Mrs. Johnson, next door, will be home all night if you need anything, but we won't be gone long."

"Okay," I said again, and went back to my window, watching the houses grow smaller and closer together as we left Lush Valley behind. Windows glowed with warm yellow light. Smoke wafted up out of chimneys. Dogs barked from fenced-in backyards. I could almost smell dinner on the stove in those houses and hear the evening news on TV. *Normal neighborhoods with normal people—families,* I thought, and just the word *families* made me sick with longing.

Chapter 5

My dad's new wife, Suzanne, was sitting on the couch watching TV when we came into the house. She muted the sound and sat up straight when she saw us.

"Hi, Fizzy," Suzanne said, smiling.

"Hi," I said to my shoes.

"I thought maybe we'd order a pizza for you before we leave. Would you like that?" Suzanne asked.

I shrugged.

Dad and Suzanne exchanged a look.

"Go put your things away," Dad ordered.

Here's another thing I hate: Every other Friday, I have to pack *and* unpack, because Dad won't let me live out of my suitcase. (Sometimes I wonder if he likes to pretend I still live here on the weekends, and if he misses me the rest of the time.) Then, every other Sunday, I have to pack and unpack again, because Mom won't let me live out of my suitcase either. This means I am either packing or unpacking about a hundred times a year. Since I'm only twelve, I have six and a half more years of this, so I'll be packing or unpacking six hundred and fifty more times—yes, I did some math voluntarily, and as much as I hate math, it's still better than packing or unpacking. I'm sick of packing.

I'm so sick of packing that if someone offered me a week on the beach, all I'd be able to think about is the packing and unpacking. No, thank you. My idea of a vacation these days is staying in one place and leaving all my stuff exactly where it is.

As usual, a pang of sadness hit me as soon as I walked into my room, which was the same as always—same rug, same furniture, same peppermint-pink-and-white-striped wallpaper Mom had chosen when I was a baby—except that there was an empty, abandoned feeling in here now. You could tell that a happy little girl had lived here once, but you could also tell that she was long gone. And she really was.

I was sitting on my bed looking over the recipes Aunt Liz had given me when Dad and Suzanne came into my room. I slid the recipe cards under my thigh to hide them.

"Pizza's on the kitchen table," Dad said.

"Thanks," I mumbled, without looking up at either of them.

"The phone number of the restaurant where we'll be is on the refrigerator."

I nodded.

Suzanne spoke up. "Fizzy, I'd love to help you with the *Southern Living* contest. Maybe we could do some cooking together tomorrow."

Aunt Liz had told my secret! My mouth fell open as I searched Suzanne for the meaning of this betrayal. But I was instantly blinded by beauty: Suzanne's shiny blond hair was pulled up, and she had on dangly earrings with red stones that matched the color of her red velvet dress perfectly.

I'd wear dangly earrings on my cooking show, too, I decided then—to match my rings.

Dad cleared his throat and gave me a look like, *Suzanne asked you a question. Answer her.* He wore a suit and tie, and his black hair was combed straight back, still wet from the shower. As it dried, it would go its own way, I knew, forming loose waves that would curl around his ears and forehead.

"Thanks anyway, Suzanne," I said, "but I'd rather do it on my own."

Suzanne pressed her red lips together so that her mouth formed a straight, lipless line.

"We'll see you later," Dad said, in a gruff voice that let me know he was *not* happy with the answer I'd given Suzanne. Then he turned and walked swiftly out of my room, taking the warm scents of soap and toothpaste with him.

I didn't care if Dad was mad, I told myself. I didn't care, and no matter what happened, I wasn't going to let Suzanne help me with the *Southern Living* Cook-Off. My dad could scream and yell and turn me over his knee, but I still wouldn't let her help me. No way. It wasn't because Suzanne's a terrible cook. She isn't. She's a terrific cook. Truthfully, she's terrific at everything she does.

Apparently, being beautiful and stylish and smart has never been enough for Suzanne. No, Suzanne has to be *perfect*. She's a lifelong straight-A student—from kindergarten all the way through college. I'll bet Suzanne never made a single B. Not one! *And* she'd been a cheerleader. As far as I could see, she'd never made any mistakes. She'd never even had a cavity!

Suzanne probably never had so much as an unkind thought about anyone—at least not until she met me. How in the world could I be expected to like a person like that? *How?*

Because meanwhile I was the weird-looking kid who was always late to school, where I made lots of mistakes, quite a few Bs—mostly in math—and stunk at kickball. I already had three fillings in my mouth and I was only twelve! And I had mean thoughts sometimes. So, see, I really *needed* to win the *Southern Living* Cook-Off and Suzanne didn't. She just didn't.

That's when I realized Suzanne proved Aunt Liz wrong. "Everybody who's anybody was nobody back when they were in school," Aunt Liz always said. But Suzanne had been *somebody* her whole life. She'd been somebody in school and she was somebody now. It didn't seem fair.

Somehow, thinking about all of this made me feel homesick again, even though I *was* home—home didn't feel homey anymore.

Mrs. Warsaw popped into my head then, and I was betting Suzanne had never seen the inside of a principal's office—or a guidance counselor's office—in her whole life! This made me feel even worse, when I really wanted to feel better.

That's why I had to cook. I *had* to do it, to feel better.

I carried my recipes downstairs to the pantry and searched the shelves for the ingredients listed on the cards. We had all the ingredients for peach pound cake. Already, I felt a little better.

I'd just put my peach pound cake into the oven and was about to set the timer when the phone rang.

I answered and Mom said, "Oh, Fizzy, I hope your dad and Suzanne won't mind—I couldn't wait to tell you!"

"If I had a cell phone, you wouldn't have to worry about that," I pointed out.

"Fizzy, I've already told you: We can't always get everything we want in life and—"

"The sooner I accept that, the better off I'll be," I finished for her.

"I'm so glad you understand," Mom said cheerfully.

"Well, anyway, Dad and Suzanne aren't here," I said.

Silence.

"Mom?"

"What do you mean, they're not there?" Mom asked, sounding much less happy than she had when I answered.

"They went out to dinner," I said.

"Without you? Fizzy, the whole point of your dad's weekend is for him to spend time with you."

Dad and Suzanne almost never went anywhere without me when I visited, so this wasn't a big deal to me. Obviously, Mom felt differently. I was tempted to ask her if she was supposed to spend every second with me when I was at her house, but I knew that'd make her mad, so I didn't.

Instead, I said, "It's a business dinner and they'll be right back."

"All right, all right—never mind that," Mom said, working herself up to sounding breathlessly happy again. "The reason I called is . . . are you ready for this, Fizzy? Are you ready?"

I gulped. "Yes, ma'am."

"Keene and I are getting married!"

I felt like my alarm clock had just gone off.

"Fizzy?"

"Um . . . yeah . . . that's great, Mom," I said, holding my stomach.

"I know! I'm so excited! Well, that's all I wanted to tell you. I have other calls to make, so . . ."

I think I said okay before I hung up. But even if I didn't, I doubted Mom had noticed.

I ran upstairs to my room, threw myself down on the bed, and let the tears come—in multiples. I gave in to them completely and sobbed. And that's how I forgot about the cake in the oven.

First, the smoke detectors went off and, great gravy, were they loud! They were a hundred times louder—and shriller—than Genghis. If that wasn't bad enough, the smoke detectors then activated something in the security system so the burglar alarm went off, too!

I ran down the stairs with my hands clamped over my ears—to keep my throbby eardrums from becoming *explodey* eardrums.

A sickly-sweet-smelling smoke filled the kitchen and drifted out to meet me at the bottom of the steps. I gagged and coughed, but before I could think what to do, the security company was calling, so the phone was ringing, and our neighbor Mrs. Johnson was banging on the front door with all her might.

· · ·

The fire department arrived just after Mrs. Johnson, who reminded me a little of Mrs. Warsaw with her tight, disapproving looks. Then Dad and Suzanne pulled into the driveway behind a big red fire truck with flashing lights.

It was just a little smoke. I mean, it's not like the kitchen was *on fire*. Nothing was on fire. *Nothing!* Still, to listen to my dad—after he learned that everything was okay—you would've thought I'd burned his house to the ground. He paced back and forth in the living room, pulling at his collar and necktie with two fingers, and called me things like "thoughtless" and "immature" and "irresponsible."

That's how I knew I'd embarrassed him. Dad wouldn't have been anywhere near this mad if Suzanne hadn't been here. He might've even laughed, but he didn't. Naturally, I took this to mean that Suzanne had no sense of humor, so she didn't allow Dad to have one either.

This made me really mad at Suzanne—Dad used to have a great sense of humor!—so I gave her squinty looks while I listened to Dad rant and rave: "Do you have any idea what happens when the bell sounds at a fire house? Grown men stop what they're doing and *run*—you put everybody to a lot of trouble, Fizzy. And furthermore . . ."

That's when I noticed that Suzanne was getting fat. Okay, well, maybe not *fat*-fat, but she definitely had a little potbelly where there used to be a perfectly flat stomach. I would've enjoyed this imperfection a lot more if Dad hadn't been yelling at me.

Dad stopped pacing. "Are you even listening to me?"

"Yes, sir," I said, giving him my full attention. Dad's ears and face and neck were a deep pink, which only confirmed what I already knew: I'd embarrassed him. He was ashamed. Of *me*. Knowing that made me feel just awful.

Dad continued, "You *will* clean up that kitchen."

"Yes, sir."

"Tonight."

"Yes, sir."

"Honestly, Fizzy, I've never seen such a mess in all my life!"

Since Dad seemed to be waiting for an answer, I offered the only one I could think of: "Aunt Liz says that a messy kitchen is a happy kitchen."

Dad's eyes bugged out of his head and the vein is his forehead bulged. "Then our kitchen is delirious. But I'm not. So go clean it up."

"Yes, sir."

I still felt so shaky and upset while I cleaned the kitchen that I put the mixer paddles and mixing bowl right into the sink and rinsed them—instead of scraping the remaining cake batter off with a spoon and eating it first—and *that* is seriously upset.

Later that night, I was lying in bed crying—just a little—into my pillow when my door opened and light from the hallway spilled into my room.

I went still, squeezed my eyes shut, and pretended to be sleeping.

I felt the mattress sink as someone sat down on the edge of my bed. Suzanne's voice said, "Growing up is hard, Fizzy."

I didn't move a muscle.

"Did I ever tell you that my mother made most of my clothes when I was growing up?"

I gave up pretending—I obviously wasn't fooling anybody—and said, "No, ma'am."

"Well, she did, and they were nice clothes. But there was nothing I wanted more than a ready-to-wear dress from the big department store downtown," Suzanne said. "My mother said those dresses were too expensive."

Is this about my clothes? Because . . . whatever, I thought, but I didn't say it. (I knew I couldn't get away with any attitude—or anything else—that night because, according to Dad, I'd almost burned the house down. Yeah, right. Did I mention that nothing was on fire? There was *no fire*.)

Suzanne continued, "Finally, for my thirteenth birthday, my parents bought me a department store dress. The first time I wore it, I spilled grape juice all down the front. I tried to clean it but I just made it worse. The stain was so bad, I thought surely it would take an entire box of soap to get it out. So, I put the dress into the washing machine, along with an entire box of laundry detergent."

"What happened?" I whispered.

"A big mess is what happened and my mother was furious. She was mad about the dress, which was ruined. She was mad that I'd wasted a whole box of laundry detergent. She was mad that I'd used her washing machine. And she was *really* mad that I'd done all this behind her back. She called me a sneak and said there was nothing worse than a sneak in her book."

"I didn't *sneak* to use the oven," I said, trying to defend myself. "I thought I was *allowed* to use the oven—I'm allowed at . . . my other house."

"That's not the point."

Then what is? I thought.

As if she could read my mind, Suzanne said, "Part of growing up is making mistakes and learning from them. You made a mistake tonight and it won't be the last. You're going to make lots of mistakes, Fizzy, and some of them are going to be painful, but it's not the end of the world."

"You made *other* mistakes?" I blurted.

"Yes," Suzanne said, and I could hear her smiling.

In a supreme moment of weakness, I heard myself say, "Could you help me make another peach pound cake tomorrow?"

"I'd love to," Suzanne said, rising from my bed.

Of course, the minute she was gone, I was sorry I'd invited Suzanne to cook with me. What the halibut was wrong with me? I lay in bed trying to figure it out, and after a while, I decided that Suzanne had broken me down with her niceness. She had "killed me with kindness." Which made me really mad. At Suzanne.

Chapter 6

On Saturday morning, Suzanne and I bundled up and went to the grocery store to buy all the ingredients for all the desserts we planned to make.

As we stood in the checkout line, a lady who knows Suzanne spotted us and rushed right over with her cart. "Did y'all have a fire last night?" she asked.

"Oh no—it was a false alarm," Suzanne said.

I shuffled my feet.

"What happened?" the lady asked, pushing her glasses up on her nose to get a better look at me.

None of your business! I wanted to say, but I remained silent and pretended to peruse the gossip magazines.

"You know how it is," Suzanne said, smiling easily. "I put some dinner rolls in the oven and then got distracted and forgot about them."

The lady nodded. "I've done that a few times myself."

"Mrs. Stein, this is Robert's daughter, Fizzy," Suzanne said.

I gave Suzanne the squinty eyes.

"Nice to meet you, Fizzy," Mrs. Stein said.

"Fizzy, this is Mrs. Stein," Suzanne said. "She just moved into the Moores' old house."

I stopped squinting. "The Moores' old house? As in *Olivia Moore*... my best friend?"

Suzanne looked a little worried, but she nodded.

Mrs. Stein looked a little worried, too—and very uncomfortable. "Thanks again for the muffin basket," she said as she took off.

I said nothing as Suzanne and I waited in line. I said nothing as we loaded our groceries into the trunk of the car. I said nothing as we drove home. I was busy thinking.

I was shocked that Olivia had moved without telling me, but I guess I shouldn't have been—we hadn't talked in months. Even so, it bothered me that I didn't know where my former best friend lived now; it seemed like proof—*more* proof—that she didn't like me anymore. I would've felt exactly the same if Olivia had been standing right in front of me, saying, "I don't like you, Fizzy Russo." *But why?* I wanted to ask her. Since I couldn't, I moved on.

To Suzanne. Why had she introduced me as "Robert's daughter"? Was I not good enough to be her daughter? Was she so ashamed of me that she wanted to make sure everyone immediately understood that I was not related to her?

We were sitting at a red light when Suzanne finally turned the heat down from her preferred setting of *Sahara Desert*—I was starting to sweat. "I give up," she said. "What's wrong, Fizzy?"

"Why did you do that?"

"Do what?"

"Why did you have to tell Mrs. Stein that I'm 'Robert's

daughter'? Why couldn't you just say 'our daughter'? Why couldn't you just let her think that we're a . . . *family*?" The word caught in my throat.

Suzanne stared at the road ahead while she thought about it. Finally, she said, "Fizzy, let me ask you a question—and I want you to answer it honestly."

"Okay," I said.

"If I had introduced you to someone as my daughter, say, yesterday, what would you have said?"

I saw Suzanne's point. "I probably would've told you that you aren't my mother and you shouldn't go around pretending you are." It sounded bad and I knew it. I knew it even before I said it, because I'd said it before. Once.

Back when Dad and Suzanne had first gotten married, I'd worried a lot about whether or not Suzanne would like me. What if she didn't? Would she be mean to me? Yell at me? Punish me? Would she load me up with chores, lock me in a dungeon, or leave me behind in the woods like the evil stepmothers in old fairy tales? I did some of this worrying out loud—apparently—because one day, Mom sort of snapped and said, "Suzanne can't do anything to you! *She* is not your mother! *I* am your mother!"

So, the next night, while Dad, Suzanne, and I were eating dinner in a fancy restaurant, when Suzanne had whispered, "Please put your napkin in your lap, Fizzy," I took the opportunity to inform her—and everyone else in the restaurant—that she was not my mother. (After that, things didn't exactly go well. I now think of it as "the worst dinner ever"—and there was nothing wrong with the food.)

Suzanne sighed. Eventually, she said, "I can't win either way, Fizzy. If I introduce you as my daughter, too, then you think I'm trying to take your mother's place, and if I introduce you only as your father's daughter, then you think I'm ashamed of you."

"Yes, ma'am, but one of those hurts worse than the other," I pointed out.

Suzanne nodded. "Yes, I see that now. I'm sorry."

My mouth twitched. It wanted to smile because I couldn't help enjoying an apology from an adult. I have to apologize to adults all the time but they almost never apologize to me—even when they're wrong. I couldn't remember the last time I'd heard an adult say "I'm sorry."

Suzanne did smile. Then she turned up the radio and started singing along. She was a pretty good singer, I noticed. *Of course she is!* I thought. *Duh!* Then I went back to being mad at her. Mostly.

But I forgot to stay mad once Suzanne and I started cooking. We made banana pudding, Derby Pie, and peach pound cake. They were all good, but the peach pound cake was so unexpectedly delicious, we decided to make another. This time, I prepared the pan with shortening and a sprinkling of sugar instead of flour. Then we added a little vanilla extract and some pecans to the cake batter. I knew it was going to be good when I licked one of the beaters I'd used on the batter.

"What are you doing?" Suzanne gasped.

"It's really good," I told her, taking another lick and closing my eyes with pleasure. "You should try it."

"Fizzy, that batter has raw eggs in it—you could get salmonella poisoning!"

I shook my head and said, "Oh no, ma'am . . . that's not true"—mostly because I didn't *want* it to be true.

But Suzanne wasn't listening. She rushed around the kitchen gathering up everything with batter in it or on it, then sent the whole stack clattering into the sink, where she turned on the hot water and reached for the dish soap.

I stared at the sink, slack-jawed, disbelieving. *Why?* I wanted to ask, the same as I would have if Suzanne had dumped all her jewelry down the drain and turned on the disposal. *Why?*

"We'd better clean up the mess before your dad sees it," Suzanne said. Then, in a split second, before I even knew what was happening, she reached over, plucked the beater from my hand, and tossed it into the sink, too.

Once I understood what had happened, my thoughts went something like this: *Unforgivable! You want to throw away* your *jewels, that's fine. But don't you dare throw away* my *jewels—or cake batter—because that is wrong . . . and unforgivable!*

I had to forgive Suzanne after the cake came out of the oven, though, because it was such an amazing cake. You couldn't really pick out the peach, vanilla, or pecan flavors; they all blended together to make one wonderfully unique flavor. Not only that, but the shortening-plus-sugar pan preparation had made the crust crunchy and sweet, like a sugar cookie.

"This is it," I said to Suzanne as we stood side by side eating warm cake over the kitchen counter.

"I know."

"But we can't really call it peach pound cake because it's something else now. What should we call it, Suzanne?"

She tilted her head from side to side, thinking. "How about Russo Pound Cake?"

"Good," I said. Okay, so even though we didn't have cake batter, we did have great cake. And pie. And pudding.

Dad must've had similar thoughts: When he came home from seeing a patient with a dental emergency and spotted all the desserts lined up on the kitchen counter, he smiled wide. First he smiled at the desserts. Then he smiled at Suzanne. Finally he smiled at me, and somehow I knew that he'd forgiven me. What can I say? Dad *loves* dessert.

Later that night, Suzanne opened the door and stuck her head in my room. "I just wanted to tell you that I had fun today. And good night."

I put my cookbook down. "Thanks. I had fun, too."

Suzanne smiled and backed into the hallway.

I called out, "Suzanne?"

She poked her head back in.

"Don't tell anybody, okay? I mean, about the contest."

Suzanne pushed the door open and stepped into my room. "Why not?"

"I don't know," I said. "I'm sort of scared, I guess."

"Scared of what?"

"It's just that if I don't win . . . if I fail . . . well, I just don't want everybody to know about it, that's all—I don't expect you to understand." And I really didn't, because Suzanne had never failed at anything in her life.

"Fizzy, everybody's scared sometimes."

"What are you scared of?" I asked.

"Snakes," Suzanne said right away. She didn't even have to think about it.

I nodded. "I'm scared of spiders."

Suzanne smiled. "See?"

I gave her a weak smile back. "Yeah, but this is different."

"I won't tell anybody, Fizzy, but keep in mind that no one can ever really succeed without risking failure."

After Suzanne was gone, I sat in bed thinking about her, thinking maybe I liked her. Did I? In the pretty toile-fabric-covered journal Aunt Liz had given me, where I keep all of my Notes to Self—so that I don't forget—I made a list of all the things I liked about Suzanne:

1) *Apologizes when she's wrong—this is HUGE!!!*
2) *Tries to comfort me when she knows I'm hurt or scared.*
3) *Tries to protect me, even if it means taking responsibility for a mistake I made, like she did at the grocery store.*
4) *Likes to cook and is good at it.*
5) *Smiles easily.*
6) *Makes Dad happy.*

Then I made a list of things I disliked about Suzanne:

1) *Washes cake batter down the sink!*
2) *Too perfect.*

That was all I could come up with. And although I found these traits highly annoying, for some strange reason, I felt willing to overlook these two—very deep—flaws. I could only assume I felt so forgiving toward Suzanne because I liked her a little. And because we had a lot of desserts downstairs.

But, I decided, she'd have to pass one more test before I could fully commit to maybe, just possibly, perhaps liking Suzanne. I got up and rooted through the old toy chest in my closet until I found it.

Then I tiptoed down the hallway to Dad and Suzanne's room. Nobody was in there, so I scurried through the bedroom and into the bathroom. Once there, I pulled my big plastic snake out from under my shirt and coiled it in the bathtub. I stepped back to admire my work. The snake looked real, all right.

I giggled all the way back to my room.

Chapter 7

I'd been so busy on Saturday that I'd mostly been able to push Mom and Keene out of my mind—repeatedly. But by Sunday morning, they were all I could think about. I never heard a word the preacher said that morning. I just stood up whenever everybody else did and bowed my head whenever Dad elbowed me.

I thought Mom knew how I felt about Keene, but then I'd never actually told her. Maybe I should, I decided. Maybe it was as simple as talking to her. I was still trying to figure out a nice, polite way to say "I don't like Keene" as I packed that afternoon, which really took the joy out of checking off my packing list—the only thing better than making a list is checking it off.

Suzanne came into my room and sat down on the bed. For a few minutes, she watched me pack. Then she said, "You've been awfully quiet today, Fizzy."

I nodded.

"Something on your mind?"

I stuffed my pajamas into my suitcase, checked them off my list, and turned to face her. "My mom's getting married." I figured the mere mention of Mom would prevent Suzanne from asking any more questions.

But it didn't. Suzanne tucked a loose strand of blond hair behind her ear and said, "That's good, isn't it? You want your mom to be happy, right?"

"Yes, ma'am . . . but . . ." I wasn't sure what to say.

"But what?"

"It's just that . . . I don't really like Keene." *And also, I can't think of a nice way to tell my mom that.*

Suzanne smiled. "You didn't like me either at first and I'm not so bad, am I?"

"No, ma'am," I said, because what else could I say?

"It's the same thing," Suzanne said.

"No, ma'am," I said, fidgeting with the button on the end of my ballpoint pen: *click, click, click.* "I'm pretty sure it's different."

"How?" Suzanne asked.

I shrugged and clicked my pen some more. I didn't know how it was different; I only knew that it was.

Late that afternoon, I stood at the back door holding on to my suitcase as Suzanne hugged me. Before she released me, she whispered into my ear, "Everything's going to be okay, Fizzy. But if it isn't, there's always a place here for you. *Always.*"

When she stepped back, I nodded my understanding. Then I said, "Um . . . I might need a cell phone—for emergencies—don't you think?"

Suzanne looked at Dad.

He shook his head.

• • •

When Dad and I were almost to the town house, I said, "Hey, I didn't forget anything this weekend."

Dad raised his eyebrows and turned to give me a look that read, *Are you sure about that?*

"What?" I said. "What did I forget?"

Dad grinned. "It'll come to you."

The sky had just begun spitting snow when Dad pulled to a stop behind Keene's parked car. I felt disappointed at the sight of that car, knowing that Keene was inside with Mom, and that I'd have to wait for him to leave before I could try talking to her.

Keene was sitting on the couch, watching reruns of *Survivor Steve*—his favorite TV show—when I walked in.

He jutted his chin in my direction—saying hello?—but never took his eyes off the TV.

For a few seconds, I stood there looking at Keene, trying to see whatever Mom sees in him. But all I saw was an ordinary guy who was extraordinarily neat: clean-shaven face, prickly buzzed brown hair that was too short to move, but plastered into place with gel anyway, wrinkle-free button-down shirt tucked tightly into the waist of crisp, creased slacks, and shoes that looked brand-new. Also I noticed that Keene has very good posture.

I thought about Dad then, how he spends most of his weekends slouching on the couch, alternately reading and watching TV, all stubbly faced, in his gray T-shirt and sweatpants.

Right away, I decided that I prefer stubble, slouchy posture, and sweatpants because they seem to invite fun and hugs—and dessert. Keene's weekend look didn't invite anything; it just said, *Hey, don't mess me up.* So I didn't.

I stayed where I was and waited for a commercial. When one came, I announced, "It's snowing," hoping that Keene would leap off our couch and hurry home, before the roads got too bad.

But he didn't. Instead, he just nodded absently and kept his eyes glued to the TV.

I gave a loud—disappointed—sigh, then slipped out of my coat, kicked off my shoes, and left them, along with my suitcase, by the front door.

I found Mom in the kitchen, fixing dinner. "Hi, sweetie," she said when she saw me. "I'd hug you but I've got hamburger all over my hands."

I nodded. "What are we having?"

"Meat loaf, peas, carrots, and mashed potatoes," Mom proudly informed me.

Yuck. Here was another reason I hated it when Keene came to dinner: I hated almost all the foods that he loved. And if I didn't do something, soon Keene would be here for dinner *every* night. Or maybe we'd be at his house. This was the thought that caused me to start feeling sick again.

I headed for the stairs.

"Don't forget your suitcase," Mom said cheerfully. "I . . . uh . . . we'd appreciate it if you didn't leave things by the front door anymore. Take your things up to your room, okay?"

By "we," I knew Mom meant Keene. I nodded, thinking, *New rules. More new rules to learn by trial and error—but mostly error.*

. . .

The sight of my room didn't help—even though the walls were painted the exact same butter-yellow that seemed so happy in Aunt Liz's kitchen. Why didn't it feel the same way here? What was I missing? I tried to figure it out—but couldn't—until Mom called me down for dinner.

When we were all seated at the dining room table, Keene lifted the platter of meat loaf and held it for me. "Help yourself, Fizzy."

I shook my head. "Guests are always served first."

Keene gave Mom an angry look and said too loudly, "I am not a guest. I am a member of this family!"

I drew back, shocked. I had only been trying to be polite. I looked at Mom for some explanation.

Mom touched Keene's shoulder and said in her softest voice, "She didn't mean it that way, honey."

Keene sort of huffed.

None of us really recovered from this little outburst for the rest of the dinner. Keene and I barely spoke while Mom chattered like crazy, trying to fill the silence.

When she ran out of things to talk about, Mom turned to me and said, "So how was your weekend?"

"Fine," I said.

"Did you remember everything this time?"

"Yes, ma'am," I said. *Except for the cake in the oven*, I thought. Dad was right: It came to me. I smiled.

Mom seemed encouraged and she smiled, too. "Fizzy, tell Keene about the *Southern Living* Cook-Off."

I dropped the smile and squirmed a little in my seat. "Oh... um... that's okay," I said, because I didn't want to tell Keene about it.

Of course, Mom told him everything anyway.

After pushing most of my meat loaf and all of my peas and carrots around on my plate—I like mashed potatoes, so I'd eaten those—for what I thought was a fair amount of time, I said, "May I please be excused?"

I half expected Mom to inspect my plate and launch into her I-worked-so-hard-on-this-meal speech, but all she said was, "You may."

I figured this meant Mom thought we'd had enough dinner difficulties for one night—I thought so, too. I picked up my plate and carried it into the kitchen.

As soon as I was out of sight, I heard Keene whisper, "Cecily, you don't really believe she can win the contest, do you? She's just a kid."

"Of course I believe she can win," Mom said.

Keene snorted.

"Fizzy's very talented—it has nothing to do with age," Mom said, sounding a little hurt.

"Oooooh-kay," Keene said, but the way he said it sounded more like *Boy, are you dumb.*

"What?" Mom said, a pinch of anger rising in her voice. "What are you trying to say?"

Get him, I thought as I stood at the counter, holding on to my dinner plate so tightly that my knuckles turned white.

"Love is truly blind, isn't it?" Keene said lightly.

I'll say, I thought, *because if it wasn't, Mom would see what a jerk you are!*

After that, in the privacy of my room, I made another list in my journal, which confirmed that I didn't like Keene—and that I had reasons—*good* reasons! It read:

1) *Prickly hair—like a porcupine!*
2) *Prickly personality—like a porcupine!*
3) *No manners. Shouldn't a guest recognize their guestness? Especially when it's pointed out to them? And shouldn't they behave like a guest?—politely?*
4) *Inconsiderate—a guest shouldn't talk about any of his hosts behind their backs—especially when I can hear him! Hosts have feelings, too!*
5) *Rude—see 3 and 4.*
6) *Doesn't believe in me.*
7) *Doesn't trust Mom enough to believe in her when she believes in me. (I'm pretty sure this indicates serious problems in their relationship.)*
8) *Gave me a hateful alarm clock for my birthday—which I'm pretty sure he received as a free gift with his* Sports Illustrated *subscription.*
9) *Re-gifts!—see 8.*

Then, to be fair, I made a list of things I liked about Keene:

1) *Practices very good hygiene.*
2) *Makes Mom happy—although I can't see how—and causes her to wear her prettiest clothes.*

This wasn't exactly enough to win me over.

When I heard Mom coming up the stairs late that night, I slipped out of bed and into the hallway.

When she saw me, Mom smiled knowingly. "It stopped snowing an hour ago and the roads are fine—school isn't canceled."

"Oh no . . . I, um . . . Is Keene gone?" I asked as she reached the top step.

"Yes." Mom reached for my hand and held it. "What is it, sweet pea? Why are you still awake?"

I begged her with my eyes, hoping she would try to understand even though she obviously felt differently, as I said, "Mom, I . . . I don't like Keene."

She sighed. "You will, Fizzy, you'll see. These things just take time."

There was something in Mom's voice that kept me from saying more. She seemed so determined, *desperate* even, for Keene and me to like each other.

Chapter 8

When the—hateful—Genghis started yelling at me on Monday morning, my room was still pitch-dark. My thoughts went something like this: *Who set that alarm? Is this some sort of joke? Because it's not funny. It's downright cruel.* Naturally I blamed Keene. Until I remembered that *I* was the one who'd set my alarm clock.

I found Mom downstairs leaning against the kitchen counter, drinking a cup of coffee.

Do you always do this? I wanted to ask, because in my opinion, a person who's always "running late" doesn't have time for leaning.

But Mom beat me to the questions. "What are you doing up so early?"

I shrugged. "I thought maybe I'd walk to school today."

Wrinkles formed on Mom's forehead. "Oh? And why is that?"

I looked down at my bare feet and mumbled, "I can't be late for school anymore."

Mom set her coffee down. "All right. Could you please explain?"

"It's just that I don't want to go to the principal's office anymore," I said softly.

Mom's eyes bulged. "When were you in the principal's office?"

"Friday... and it wasn't the first time," I told her. "I'm late a lot, Mom." I felt bad saying it. I really did.

Mom's face melted into a sad sort of smile. "I'm still trying to figure out how I can do it all... by myself... and I... I..." She shook her head and then showed me her palms. "All right."

I headed for the stairs but when I glanced back, Mom looked so sad. "Enjoy your coffee," I tried.

Another sad—guilt-loaded—smile.

I felt bad for her and didn't want to leave her like that. I took another step toward the stairs and stopped, grinning as the idea hit me. I turned to face Mom fully as I said, "You know, I'm sure you'd feel much better about all this walking if I had a cell phone."

It worked: Immediately Mom hardened and said, "We can't always get everything we want in life, Fizzy."

I took the stairs two at a time and was already at the top when she called after me, "The sooner you accept that, the better off you'll be!"

Mom caught me at the front door on my way out and said, "I almost forgot: I spoke to your aunt Liz and she wants you to plan on going over to her house after school every day this week to try out a bunch of recipes for the contest."

I should've known something was up right then because Mom had talked to Aunt Liz—which she hardly ever does anymore—but I didn't. So I just said, "Okay . . . but what about Thursday?"

"What about Thursday?"

"It's Parents' Night at school," I reminded her.

"Oh, right. Keene and I will pick you up from your aunt Liz's right after work," Mom said.

"Why?" was the word that tumbled out of my mouth, as in, *Why would Keene come?* He wasn't my parent. He wasn't anybody's parent.

"For Parents' Night—isn't that what we're talking about?" Mom gave me a quick hug and headed for the stairs before I could think what to say.

It was a cold, gray morning. A few patches of snow remained in shady spots, with dead grass sticking up through them in an unruly way, which seemed out of place in Lush Valley, where yards aren't only strictly ruled but sculpted—some bushes are even carved into shapes, like art—even in the winter.

As I walked, I thought about how happy Coach Bryant would be to know that I get all this fresh air and exercise outside of gym class. I thought maybe I should tell him. Maybe he would say something like, *In that case, you're excused from gym class for the rest of the year. Sit and read all you want. You've earned it, Fissy.* (The way Coach Bryant says "Fizzy" sounds more like "Fissy," which sounds a lot like "Fussy" to me. I don't like it.)

That's what I was thinking when I heard a door slam somewhere behind me. I turned to look: Zach Mabry, a blond-haired, blue-eyed boy who was in my math class came thudding down his front steps, hugging an unzipped black backpack with papers sticking out of the top, his coat half on and half off. When he saw me looking at him, he showed me his teeth.

Was that supposed to be a smile? I wondered as I turned back around and resumed walking.

Within two minutes, I heard footsteps on the sidewalk right behind me. I glanced over my shoulder.

Zach showed me his teeth again.

I decided not to look behind me anymore and picked up my pace.

But Zach stayed right with me. Until we reached the school. Then he jogged around me, moving ahead of me as he hurried for the door. But he didn't open it. Instead he just stood beside it, staring at me.

I slowed, looking around, unsure. But then I remembered how I couldn't be late again.

As I neared the door, Zach pulled it open and held it for me.

"Th-thank you," I stuttered as I passed.

"It's not working, is it?" Zach asked as he entered right on my heels.

I turned. "What?"

"My I'm-not-a-violent-maniac smile."

"Oh," I said. "Um . . . why do you need an I'm-not-a-violent-maniac smile?" *Because wouldn't only a violent maniac need one of those?*

"When you come up behind kids who've lived on the street, especially girls...," Zach started, but then he shook his head and said, "never mind." He smiled a real smile, winked, and added, "Have a nice day." I was a little unnerved. Until I reminded myself how cool I must look as a walker, *voluntarily* showing up for school and all.

Mr. Moss began science class by announcing that he was going to give each of us a marble for our experiment in force of motion, and then spent the next twenty minutes detailing all the terrible things that would happen if we lost our marble. For starters, we wouldn't get another one and we wouldn't be allowed to share with a friend. *Blah-blah-blah.* Cut to catastrophe: "Accidentally drop your marble down the sink? Too bad. No marble means no experiment, no data, and no points—you'll get a zero if you lose your marble. Everybody understand?" We all nodded, ready to get going.

But Mr. Moss spent another ten minutes telling us about all the different kinds of marbles he'd collected over the years: cat's eyes, devil's eyes, rubies, butterflies, bumblebees, and so on.

When we—finally!—got our marbles, I saw the one belonging to the girl in front of me, Miyoko Hoshi, roll off her table and onto the floor. The girl at the table beside Miyoko's, Ada Montgomery-Asher—one of Buffy's friends—saw it, too, because Ada reached out with her foot, rolled Miyoko's marble under her table, and kept it there, under her shoe. Miyoko must not have noticed, though, because she didn't move.

I knew that losing her marble would upset Miyoko even though I didn't really know her, because it's obvious to everyone that she's a very serious student. In addition to science, we also have math and gym/health class together. Miyoko's not much better than I am at gym, but she tries a lot harder—and never forgets her gym shoes. Also she sometimes asks health questions that stump Coach Bryant—he always says, "Uh . . . I'll have to get back to you on that."

When Mr. Moss stopped talking, the room suddenly fell silent and everyone looked up at him. "Well?" he said. "What are you waiting for? We only have ten more minutes! Get to work!" As if *he'd* been waiting for *us* to stop talking this whole time. Right.

Miyoko moved things around on her table and bent to look underneath it, while Ada worked on her experiment and paid no attention to Miyoko.

I tapped Miyoko on the shoulder to get her attention and then crawled under Ada's table and retrieved the marble from under her bedazzled shoe.

Ada gave me a surprised look, as if to say, *Oh, I had no idea that was there!*

I gave her a squinty look back to let her know that I knew better.

I placed the marble on the table in front of Miyoko, who looked at me like I'd just placed Julia Child's famous beef bourguignon in front of her when she was starving to death—I thought she might cry. Since I'm against tears at school, I immediately dropped Miyoko's gaze and hurried back to my table.

• • •

Miyoko sat beside me that afternoon during our health lesson, which was only fifteen minutes long, because Coach Bryant thought fresh air and exercise were way better for us than sitting around in the dank locker room where he had to teach health. We spent the remainder of our class time outside, but at least we were free to do what we wanted. I'd brought my cookbook—and my coat—just in case.

It wasn't as cold as it had been this morning, but the sky was still the color of oysters, which made me hope for more snow. I sat down with my book beneath the big sugar maple that had turned bright candy-apple red back in October. I love candy apples. *Hey, maybe Aunt Liz and I can make some this afternoon, just for fun,* I thought.

Miyoko wandered over then and squatted beside me on the grass—she'd brought her coat to health class, too, I noticed. Miyoko had the same kind of dark, shiny hair as Aunt Liz—except that it was stick straight—with almond eyes and a perfectly straight nose. She was pretty, maybe the prettiest girl in the whole sixth grade. I smiled at her.

Miyoko smiled back.

"I'm Fizzy," I said.

Miyoko nodded. "Thanks for this morning, Fizzy—with the marble."

"Sure. Hey, can I ask you something?"

Miyoko nodded again.

"Does Coach Bryant ever get back to you on your health questions?"

"Never."

"That's what I thought."

"But he always gives me an A in health," Miyoko offered.

"Well, he kind of *has* to, doesn't he?" I said. "I mean, you know more than he does."

"I think most people do, don't you?" Miyoko said.

We both laughed.

After that, despite the fact that my behind was going numb from sitting on the cold ground, I felt something digging into it. I thought it was probably a rock and stood up to look. But there wasn't any rock. I felt my back pockets. There was something in one of them. I reached into it and pulled out a gigantic black spider.

Now, even though I should've known the spider wasn't real—and I sort of did—I threw it down like it was the deadliest spider known to man.

Miyoko gasped and sprung up to a standing position.

I laughed a nervous little laugh and picked up the spider, to show her that it was fake.

Miyoko laughed, too. "What? . . . Why? . . ."

I shook my head. "I have no idea."

At the end of the day, I found Miyoko waiting near my locker. "Do you like spiders?" she asked seriously, like this was the most important question in life.

"No way," I said.

Miyoko appeared relieved. "Me neither."

"Why?"

"Why don't I like them?" Miyoko asked, as if this was pure crazy talk.

"No, why do you care whether or not *I* like them?" I clarified.

Miyoko shrugged. "Oh, I don't . . . I just . . . well . . . I'm not sure I could be friends with a person who likes spiders."

I smiled then. Because we were friends.

As I walked to Aunt Liz's house, I tried to figure out how that spider might have gotten into my jeans. But I couldn't. So I tried to remember the last time I'd worn them. It had been Saturday. It was only then that I remembered the snake I'd left in Suzanne's bathtub. So not only did Suzanne know how to take a joke, she knew how to play one! She *did* have a sense of humor! I laughed out loud. At that moment, I fully committed myself to maybe, just possibly, perhaps, liking Suzanne, instead of working so hard to stay mad at her all the time.

Chapter 9

Aunt Liz already had her apron on and was ready to get cooking as soon as I arrived that afternoon—well, almost. I followed her into the kitchen, where she plucked a garbage-bag tie out of a drawer and used it to pull up her hair in about three seconds flat, after which Aunt Liz looked like she'd just come from the salon, where the chicest updo ever had been created on her head. What can I tell you? Aunt Liz is the only person I know who can make something so elegant using a lowly garbage-bag tie—I wished I had some garbage-bag ties at home but we buy drawstring bags—and somehow I knew I'd never master this technique, not with all the garbage-bag ties in the world.

Anyway, after Aunt Liz performed her hair magic, we turned on the radio and sang and danced, cooked and tasted, and laughed and laughed. I was as happy as a birthday cake, right up until Aunt Liz said, "Listen, Fizzy, when I called your house to make plans to cook with you this week, your mom asked me to talk to you about Keene."

I froze right where I stood.

Aunt Liz put an arm around my shoulders and gently guided me to the kitchen table.

When were both seated, she said, "What is it, Fizzy? Why don't you like Keene?"

I still didn't want to tell Aunt Liz, but I knew I had to now. Because Mom was so determined for Keene and me to like each other, she had broken one of the A.D. Rules. As far as I can tell, the A.D. Rules are these:

1) Mom and Dad don't speak to each other. If they absolutely have to communicate—about me or my schedule—they do it by email or text. If they end up on the phone with each other—while trying to reach me—they say "one moment, please" and give me the phone, or they take messages in a polite way, like they're speaking to someone they've never met. (Occasionally, they speak of each other to me, but they never say the other's name and they always put a "your" in front of it, like "*Your* mom wants you home at two o'clock," or "*Your* dad wants you to call him," as if to say, *Don't forget: These are your people. Not mine. Yours. All yours.*)
2) Mom and Dad don't see each other. Whenever they're in the same place at the same time, like at a school play, or even in front of their own houses when I am being dropped off or picked up, they each pretend not to see the other. There is no waving, no smiling, no nodding, no nothing. (I've decided they think waving, smiling, and nodding are

too friendly—or maybe too forgiving. They do not feel friendly or forgiving toward each other. At all.)

3) All of these same rules apply to most of the extended family, which is to say that Dad's side of the family no longer exists as far as Mom's side of the family is concerned, and vice versa. Except for Aunt Liz, who doesn't seem to know any of the new A.D. Rules. Also, sometimes Mom waves at Aunt Liz—but only sometimes—when she's in a good mood. (Mom and Aunt Liz used to be close.)

4) While I am allowed to speak to or see any family member almost anytime, it's best if I play along according to the above rules. In other words, it's best if I don't say too much about the side of the family that no longer exists to whomever I'm with. So, usually, it's easiest if I don't say too much of anything at all. Otherwise, I have to think—and carefully sift my thoughts—before I speak. For example, I can say to Mom matter-of-factly, "We went to the zoo this weekend." But it'd be a mistake for me to gush, "Dad took Suzanne and me to the zoo and he bought me a stuffed leopard!" There is a world of difference between those two sentences. One is acceptable; the other will ruin the rest of the day.

Of course, nobody's ever actually explained these A.D. Rules to me, but they exist as surely as both sides of my family, believe me. So Mom actually speaking to Aunt Liz about personal stuff meant that Mom was desperate. Desperate to marry Keene, desperate for everyone to be okay with it, just desperate. I knew then that Mom was probably going to marry Keene Adams. And it made me . . . *angry*.

I crossed my arms over my chest. "Do you think I'm smart?"

"Of course," Aunt Liz said.

I lifted my chin. "Keene doesn't think so."

"How do you know?"

I hesitated. Then I lowered my head and confessed, "I can hear everything that goes on in the living room through the vent in my bedroom."

"I see," Aunt Liz said evenly, and somehow I knew she was trying not to react. When I looked up, Aunt Liz's lips were all puckery but her eyes were laughing.

I felt a little encouraged, so I explained, "After Mom introduced Keene and me for the first time, I went upstairs to my room."

"And?"

"I could hear Mom through the vent, trying to convince Keene what a great kid I am—like I heard her say, 'She's really smart,' and stuff like that."

"What did Keene say?" Aunt Liz asked.

"He said that *all* parents think their kids are smart, but if they really were, we'd have a world full of geniuses."

"I see," Aunt Liz said. "Well, there's some truth in that, Fizzy, but it doesn't mean you aren't smart. It just means that Keene will have to decide that for himself. And I'm sure he will."

I bit my lip.

"Anything else?"

"He doesn't believe I can win the *Southern Living* Cook-Off," I said.

"That's not going to be a problem, since you *are* going to win," Aunt Liz said, smiling.

I shook my head. "Also, Keene doesn't want my mom to love me anymore . . . so she's probably going to stop soon."

That's when I think Aunt Liz started to get it because she frowned and said, "I need you to explain that to me, Fizzy."

I uncrossed my arms, leaned across the table, and whispered, "I once heard Keene ask Mom if she loves me more than she loves him."

"What did your mom say?"

I sat back and shrugged. "She just said that I'm her daughter and he'd understand when he had a child."

"That's true," Aunt Liz said. "Did Keene say anything else?"

"He lowered his voice so I couldn't make out the words. But I could tell from his tone that he was sort of . . . I don't know . . . *mad* maybe—he wasn't happy, that's for sure."

Aunt Liz scowled for a split second, but I saw it. Then sleet began pinging the windows and she twisted around to look. By the time Aunt Liz returned her attention to me, she was wearing her pleasant little smile again.

I tried to give her a pleasant smile back but, honestly, I wasn't feeling all that pleasant anymore.

• • •

Because of the sleet, and because Mom was running late, I stayed and ate dinner with Aunt Liz and Uncle Preston. "Dinner" was a variety of completely unrelated foods that Aunt Liz and I had cooked that afternoon. But leave it to Aunt Liz to pull any meal together with the perfect theme—or a garbage-bag tie.

As soon as Uncle Preston arrived home from work, Aunt Liz announced, "We're having an Around-the-World tasting menu tonight, courtesy of the *Southern Living* Cook-Off!" Uncle Preston looked very impressed, and he's not an easy guy to impress—he's *literally* been around the world many times on his business trips.

We'd just started in on the desserts—a French apple tart and a German combination of ice cream and hot fruit sauce called *Eis und Heiss*, which means "ice and hot"—when the doorbell rang.

Aunt Liz went to answer it while I put on my coat and grabbed my backpack. Uncle Preston stayed with the desserts.

When I stepped out onto the front porch, Mom said, "Hi, sweet pea. Go on out to the car. I'll be right there."

I did as I was told, while Mom and Aunt Liz stood talking under the cover of the front porch, hugging themselves and rubbing their arms against the cold.

When Mom finally came rushing out and got into the car, she didn't look at me, and I could tell she was sort of upset. She slammed the door shut, shook the icy rain off her coat in an agitated way, and put on her seat belt. Then she backed out of the driveway and started home, all without looking at me—*still*.

That's when I thought maybe Aunt Liz really had understood. Maybe Mom understood now, too. Maybe she wouldn't marry Keene after all.

I watched the windshield wipers scrape back and forth in rhythm.

When we were almost home, I said in a voice barely loud enough to be heard, "Mom, I don't want Keene to come to Parents' Night."

"All right, all right," Mom said, as if I'd told her this forty-three times already, when I hadn't even said it once.

Mom parked, pulled the keys from the ignition, and turned to me. "Fizzy, you are my daughter. That means I'm always going to love you. Always. No matter what. Do you understand?"

I shrugged.

Mom continued, "There's nothing you or anyone else could ever do that would make me love you any less."

"Okay," I said.

"Okay?" Mom said.

I nodded. "Okay."

I did feel a little better. I felt like I'd been heard. And maybe even understood.

Maybe.

Chapter 10

The sun was shining on Thursday morning and the sleet was gone, which improved my mood if not the temperature—still cold. As soon as I reached the top of the hill on Dahlia Drive, Zach Mabry jumped off his porch swing and started waving like crazy.

I looked around but didn't see anybody else, so I raised one hand in a weak wave, so as not to be rude—just in case he was waving at me.

Zach pounded down the steps to the end of his walkway and waited. *For me?* I wondered. He'd abandoned his usual jeans and lace-up boots, in favor of khaki pants and loafers.

When I was close enough to hear him, Zach said, "Mind if I walk with you?"

I shrugged. "No, I don't mind."

"I'm Zach," he said as he fell into step beside me.

"I know," I said. Even though, like me, Zach was new to Lush Valley Middle School this year, unlike me, everybody knew who he was because of Buffy Lawson's crush on him.

"You're Fizzy," Zach said.

"Yep, I know that, too," I said.

Zach chuckled.

I glanced over at him. Buffy was right to like him: Zach was cute with his messy blond hair, icy-blue eyes, and lopsided grin.

"You sure do walk to school early," Zach commented.

"I can't be late," I said. *Again,* I thought.

Zach nodded and we walked the rest of the block in silence.

Again, he held the door open for me at school, and again, I said, "Thank you," as I passed through it.

Zach hurried into the building behind me. "Come with me," he said. "This way."

I looked up at the clock in the hallway.

"We've got time. C'mon," Zach insisted.

I followed him into the cafeteria, where some kids arrived early and ate breakfast, up the center aisle, and to the counter.

"Good morning, Mrs. Hunt. You sure do look nice today—I think you're the only woman I know who can really pull off a hairnet," Zach told the lunch lady.

Mrs. Hunt smiled at him.

"This is my friend Fizzy," Zach told her.

"Hello, Fizzy," Mrs. Hunt said.

"Hi, Mrs. Hunt. It's nice to meet you," I said.

Mrs. Hunt held up her index finger as in, *Just a minute*, and disappeared into the back.

I looked at Zach.

"Just wait," he whispered.

Mrs. Hunt returned with two steaming hot chocolates topped with mini-marshmallows in Styrofoam cups and handed them to Zach. "Y'all go on now. Hot chocolate's supposed to be for faculty only."

"Thanks," Zach said.

Mrs. Hunt smiled, nodded, and said again, "Y'all go on now."

"Wow," I said to Zach as soon as we were out of earshot.

He grinned his lopsided grin. "Imagine what I could do in cool clothes."

"Cool clothes?" I repeated.

"Yeah . . . like black leather. I probably wouldn't even have to talk if I was wearing black leather—because my clothes would tell everybody how cool I am."

I laughed. "If that's true, then I need some black leather, too—no one around here seems to recognize my coolness."

"I do," Zach said.

I smiled, certain that was because of my new walker status.

Lush Valley has more of everything, so naturally Lush Valley Middle School has more of everything, too. Where my old school just served lunch, LVMS serves breakfast *and* lunch, and where my old school had only one picture day per year, LVMS has *two* picture days: one in September and one in January. But I'd stopped paying attention to the announcements about picture day at school years ago. What difference does it make? No outfit is going to hide my freckles.

But nobody else felt that way, apparently. So today everyone looked very . . . matchy-matchy for pictures. Even Zach Mabry in his khakis and loafers. Even Miyoko, who wore a sweater set with a plaid skirt, matching plaid headband, and shoes with little plaid bows on the toes. It was the shoes that were a problem.

When Miyoko reported to gym class, Coach Bryant took one look at her shoes, shook his head, and said, "If that's all you've got, you won't be able to play kickball with us today, Meryoko."

Does he ever get anybody's name right? I wondered.

Miyoko turned and looked at me with pleading eyes, as if she hoped I could yank gym shoes out from under some other girl's feet—like her ruby marble.

"I forgot my gym shoes, too," I said to Coach Bryant.

Coach Bryant didn't look surprised. "Fissy, how is it that you always remember your coat and your book, but never your gym shoes?"

I didn't exactly have an answer handy.

Coach Bryant opened his mouth to say something more—probably about the reports Miyoko and I owed him—but when Miyoko sniffed, he closed it again.

We both looked at Miyoko, who hung her head and sniffed again.

"Oh . . . no . . . don't . . . uh—" Coach Bryant stammered. Then he looked at me like, *Help.*

"Maybe she just needs some fresh air," I suggested.

"Yeah," Coach Bryant immediately agreed. "Come on outside with us and get some fresh air at least."

Now, Coach Bryant couldn't very well take "Meryoko" outside and send "Fissy" to the library, could he? I mean, that wouldn't be fair.

Miyoko and I were headed for our candy-apple tree when Buffy started snickering with her friends and I heard Christine say, "Miyoko."

I was going to ignore them but Miyoko stopped immediately and turned to face the girls.

They all stopped what they were doing, too, and looked at her like, *What?*

Suddenly, Miyoko's hands chopped through the air. "Hiiii-yaaaah!" she shouted. Then she did a little kicky thing.

My eyes practically popped out of my head. I could hardly believe what they were telling me. Was pretty little Miyoko Hoshi about to hurt somebody? I could tell that Buffy and her followers were wondering the same thing. They all went completely silent and still—except for their shifty, nervous eyes and a couple of gulps.

Miyoko turned away from them and walked toward me.

When we reached the tree, I whispered, "Do you know karate or something?"

"No," Miyoko said, "but I know how to *pretend* I know karate."

I burst out laughing. Then Miyoko did, too. We both fell all over ourselves laughing.

When we began to settle, I said, "Maybe you could teach me some pretend-karate." I had lots of uses for pretend-karate: at school, at home . . . well, okay, it would only work once at home, because Mom would tell Keene that I didn't actually know any karate . . . unless I *did*. Maybe I could take real karate lessons!

Chapter 11

Aunt Liz and I were the ones running late on Thursday evening. It turned out that making individual cheese soufflés—a possibility for the Party Starters category of the cook-off—was a little more time-consuming, complicated, and difficult than we'd thought. We'd stirred and whipped and beaten our hearts out. We'd even made little tinfoil collars for our soufflés, to keep their heads from spilling over and running down the sides of their cups. And when we finally put them in the oven, we'd kept a close watch. They'd risen to form perfect little golden peaks. So we pulled them out of the oven. Right away, the peaks sank back down into the cups, even as I commanded them, "No, no, no, no, no—don't do that!"

Aunt Liz gave me a sympathetic look.

"Can we put them back in the oven?" I asked her.

"Afraid not. They're done for. We'll have to start over tomorrow."

My heart sank soufflé style as a car horn honked twice—*beep! beep!*—outside.

For once, I was glad that Mom had been running late, too—because she doesn't like to hang around Aunt Liz's house waiting for me. We went straight to school.

I left Mom at my homeroom door and headed for the gym, where all the students were gathering. I skittered past our music teacher, Mrs. Gita, before she could see me—and place me—and placed myself next to Miyoko on one of the three risers.

Miyoko smiled and said, "She's not going to let you stay here—you're too tall."

"We'll see," I said. Then I scanned faces, looking for Zach. He wasn't there yet.

Zach was the last student Mrs. Gita placed on the risers. When she stepped back to look, I bent my knees to make myself the same height at Miyoko.

It worked. Soon the gym was filled with singing. Once, Zach caught me staring at him, but I looked away—quick. Twice, we practiced the songs we were going to sing for our parents to end Parents' Night. Then parents started showing up.

Now, the best part of the actual performance—for me at least—was when Buffy Lawson fell off her riser. I mean, one minute she was standing there singing, and then *SPLAT!* She was on the floor! I didn't dare look at Miyoko, but I grabbed her hand and squeezed like, *Great gravy!* She squeezed back like, *I know!*

A few teachers and other adults rushed forward to see to Buffy, while Mrs. Gita's hands continued dancing up and down as she stood in front of us. *Keep singing,* she mouthed. *Keep singing!* So we did.

Mrs. Sloan—the gypsy guidance counselor—helped Buffy to her feet just as our last song ended. The gym exploded in applause. I'm pretty sure Buffy thought the applause was for her because she smiled a shy smile and waved at the audience. Yeah, right. I mean, when you get up out of bed in the morning, do people *clap* for you? No, because let's face it: The ability to stand up isn't exactly awe inspiring.

I'd just said good-bye to Miyoko and her parents and was looking for Mom when Christine Cash came up to me chewing pink bubble gum like it was her only purpose in life. (I'm not allowed to chew gum because Mom says it isn't ladylike.)

Christine said, "Is it true that Miyoko Hoshi is a black belt in karate?" *Chaw. Chaw. Chaw.*

I started to smile but caught myself, and instead met her eyes with my own very serious ones. "Yeah," I said. "She knows three ways to kill a grown man instantly with her bare hands—they're like . . . *weapons*."

Christine's eyes flew open wide and she gasped. Then she had a little coughing fit—she'd nearly choked on her bubble gum.

"Um, are you okay?" I asked.

"Fine," she said before she scurried away.

Note to self: Chewing gum is not only unattractive, it's dangerous!

I spotted Mom by the piano, talking with Mrs. Gita. I made my way over to them and then wished I hadn't. Mom was trying to sell Mrs. Gita advertising in the newspaper!

Here's the thing: It was Mom's job to sell advertising in the newspaper and that was fine. The problem was that she was *always* trying to sell advertising, even when she wasn't at work. Whenever Mom met somebody new, her first question was where they worked and how the company advertised. It was pretty embarrassing.

I gave Mom a look like, *Please stop*.

"Fizzy, how would you like to take piano lessons?!" Mom said enthusiastically, as if she were offering me a once-in-a-lifetime opportunity.

"Um . . . I don't know," I said, giving Mrs. Gita an apologetic smile.

"We'll discuss it—I'm sure Fizzy would love piano lessons," Mom told Mrs. Gita.

Someone tapped me on the shoulder. I turned.

"Hi," Zach said.

"Oh, hi." I felt fluttery feelings in my stomach, but not like sickness—like something else. Then I realized Mom was staring at us. "Mom, this is my friend Zach Mabry. Zach, this is my mom."

"A pleasure to meet you, Mrs. Russo," Zach said. "I can see where Fizzy gets her good looks."

My mouth fell open. *Good looks? He thinks I have good looks?*

Mom must've been thinking the same thing because she raised one—very suspicious—eyebrow at Zach.

He showed her his teeth.

"Well, we'd better get going," Mom said to me, and then she said to Zach, "It was very . . . *interesting* meeting you."

"You too," Zach said to Mom's back.

● ● ●

It had turned dark outside and the wind stung my face with cold. Mom and I ducked our heads and hurried to the car.

As soon as we were inside with the doors shut, I said, "I don't want to take piano lessons."

Mom ignored me and turned the heat on full blast. As she backed out of the parking space, she said, "Mrs. Gita is thinking of advertising piano lessons with us, so we're thinking of taking piano lessons with her—that's the way the world works."

"Well, *I* was thinking of taking karate lessons," I said, feeling very... *kicky*.

No answer.

Just to be clear, I added, "I *definitely* want to learn karate."

Chapter 12

Mom and I stopped for dinner at Lush Valley Bistro. When we entered the restaurant, the hostess looked us up and down and seemed unimpressed, probably by our lack of designer stuff, but she gave us a table—in the back. After the server brought our drinks, took our orders and our menus, Mom and I were quiet for a few minutes.

Then Mom said, "What are you thinking about, Fizzy?"

I was thinking that Keene must've bought a lot of advertising from Mom because she was never trying to sell to him the way she was always trying to sell to other people. But I knew Keene bought a lot of advertising. I'd seen the ads for his hardware store—they were in the newspaper every day and they were big and colorful. That's how Mom met Keene to begin with: He bought advertising from her.

"Fizzy?" Mom said, growing impatient.

"I'm sorry," I said. "What did you say?"

Mom shook her head and smiled. "Nothing. Never mind. So. Zach Mabry."

"What about him?"

"He's a little *slick*, don't you think?" The way Mom said

"slick" was the way she might've said "slimy." So I knew it wasn't a compliment.

I shrugged.

"Is he your boyfriend?"

"No, ma'am, just a friend."

Mom didn't look like she believed me.

"He's *just a friend*," I said emphatically, but what I thought was, *You don't like my friend Zach? Well, I don't like your friend Keene. So we're even.*

"All right, all right," Mom said, showing me her palms. "Listen, I want to talk to you about the wedding."

The wedding? There was still going to be a wedding? "Um . . . okay."

"Keene and I want to involve you, honey, because our wedding isn't just the usual joining of two people, you know."

"It isn't?"

"No," Mom said. "It's the joining of a family, one that includes you."

The queasy feeling I got told me that wasn't true, but even so, I said, "Okay."

Mom smiled brightly. "I want you to be my maid of honor, Fizzy. I want you to stand up in the front of the church with me."

"Ummm . . . okay . . . I guess."

"And I was thinking maybe we could go shopping for dresses on Saturday morning."

I nodded.

"I want you to choose your own dress and help me choose

mine," Mom said. "Oh, and I was thinking you might like to have your own cake."

"Cake?" I sat up a little straighter in the booth. There was nothing wrong with cake. I mean, cake is always good, right?

Our dinner arrived, and after the server set our plates down and disappeared, Mom announced, "There will be wedding cake and groom's cake and Fizzy's cake."

I took a bite of a French fry. "Okay," I said. "I know exactly the cake I want."

Mom clapped her hands together merrily and said, "Wonderful! Tell me."

"I saw a picture of this cake in *Southern Living*. It has three square tiers and pale purple icing, with tiny deep-purple violets all over it, and . . . well, it's just the prettiest cake I've ever seen."

Mom frowned. "Purple? You want a *purple* cake?"

I nodded and popped another French fry into my mouth.

"But my colors are peach and cream," Mom said. "Everything for the wedding is going to be done in shades of peach and cream."

I didn't really see how that was a problem myself. I mean, we were talking about cake, not curtains.

"A purple cake won't match anything, Fizzy," Mom said, still not touching her salad.

"*Qui se soucie?*" I said, which is French for "Who cares?"

Mom stiffened. "Fizzy, you know I think it's rude when you speak French."

"Then maybe you shouldn't have moved me to Lush Valley—they didn't teach French at my old school."

"No, I'm glad you're learning French; I just think it's rude to speak it to someone you know doesn't understand—it's like whispering in front of someone you know can't hear you."

I didn't respond.

"As for the cake, I don't know what to think of a purple cake," Mom said. "And no one else will know what to think either."

Suddenly I was mad. I'd had enough and I was just plain mad. I sighed loudly and said, "They'll think you did something nice for your daughter. They'll think you let her choose. For once!"

Mom's eyes narrowed. "For once? *For once?*"

Now, if I was really as smart as Mom thought I was, I would've stopped talking. But I was mad, so I didn't. Instead I said, "Yes, for once, Cecily." (My mom hates it when I call her by her name—it's way worse than speaking French.)

Cecily crossed her arms over her chest and her cheeks turned pink.

I continued, "I never get to choose, *never*! I didn't choose you and I didn't choose Dad. I didn't choose for you to get divorced. I didn't choose who I was going to live with. I didn't choose Lush Valley or our town house or my school, or even piano lessons, and I surely didn't choose Keene Adams to be my new stepfather!"

Our server appeared out of nowhere to ask how everything tasted. Mom smiled easily and said that everything was fine. I almost believed her, but when our server walked away, she took Mom's smile with her.

Then, through clenched teeth, Mom said, "Close your mouth and eat your dinner, Fizzy."

Now, just how was I supposed to do that?

• • •

I should've been sleeping, but I was still up doing my homework when Mom came into my bedroom that night, wearing pajamas with a cardigan sweater. "You've been up late every night this week, Fizzy."

It was true. Since I'd been cooking with Aunt Liz all afternoon every day, it had been late when I started my homework each night. And I had a lot of homework—like I said, there's more of everything in Lush Valley, even homework.

"My book report's due tomorrow," I said, without looking up from my paper.

Mom sat down on my bed. "You know, Fizzy, pretty soon, you're going to be all grown up and you're going to go off to college."

"Culinary school," I corrected.

Mom smiled a sad little half smile. "The point is that one day you're going to be gone, and I don't want to be alone for the rest of my life. I want a family."

Me too. I want a family, too, I thought, but I didn't say it. Instead I put my pencil down, got up from my desk, and went to sit beside Mom. "I'm your . . . it—*I'm it*."

"Yes, and you'll always be my family," Mom said. "But one day, you're going to grow up and set out into the world to create your own life, your own home, your own family."

I stared into my lap and stammered, "So you want Keene to be your . . . f-family."

"Yes," Mom said, but the way she said it was like, *Yes and . . .* I'd heard the *and* even though she hadn't said it.

I tried to think. "Do you want more children?" I asked.

"I think I do," Mom said, taking my hand in hers.

"So I'm not enough," I whispered as tears burned behind my eyes.

"You are wonderful," Mom said. "*You* are what makes me want more children—I'd like to have three more just like you."

"*Three?*" I felt sick.

Mom smiled. "Yes, but I'll take just one. One doesn't sound so bad, does it?"

"I guess not," I said, even though it all sounded pretty bad to me.

"Fizzy, look at me," Mom said.

I lifted my head and met her soft green eyes.

"Can you imagine giving up on your dream of becoming a chef?"

I swallowed. "No, ma'am."

Mom nodded. "I can't give up on my dream of having a family either." She stood.

I just sat there.

Mom placed a gentle hand under my chin and bent to kiss my cheek. "Good night, sweet pea—oh, and I promise to think about the cake."

When she was almost to the door, I said, "A very wise woman once told me that things don't matter, and what other people think about our things certainly doesn't matter. People are what matter."

Mom stopped moving but didn't turn around. She sighed. "Fine. You'll have your purple cake."

Chapter 13

On Friday morning, I found Zach waiting for me on his front porch again. "Your mom doesn't like me," he said as soon as he met me on the sidewalk.

I turned away, watching my breath crystallize on the air like smoke. "C'mon," I said. "It's too cold to stand around." We started walking.

"What'd she say?" Zach asked.

I thought about what Mom had said and felt myself smile. Zach would probably take it as a compliment—black leather and all. "She called you 'slick.'"

Zach laughed. And laughed. And laughed.

So I did, too.

We were almost to school by the time we settled down. As we stepped off the sidewalk, into the grass, Zach said, "Your mom's right, though. When you're on your own, you learn real quick that it's best for everybody if you just say whatever the adults want to hear, you know?" And then he pulled the door open for me.

I nodded as I passed, like I completely understood, even though I wasn't sure what Zach was talking about. All I really knew was that maybe he didn't think I had good looks after all—he'd just said what he thought Mom wanted to hear—right?

"Later," Zach said.

"No hot chocolate?" I half whined, coming to a dead stop in the hallway, causing the boy behind me to bump my backpack.

"Can't do it every day—wouldn't want to take advantage." Zach smiled and winked, then headed for his homeroom.

I thought about what Zach had said all through science class. And even though I didn't fully understand where he was coming from, I was pretty sure I understood part of it: It really was easier on everybody to just say whatever the adults wanted to hear, or, in my case, it was easier on everybody if I *didn't* say whatever the adults *didn't* want to hear—like how Mom doesn't want to hear about Dad, and Dad doesn't want to hear about Mom. Would I become "slick"? I tried to imagine myself dressed in black leather from head to toe, maybe with silver, spiky bracelets around my wrists, which caused me to giggle.

"Something you'd like to share, Miss Russo?" Mr. Moss said.

I sat up straight. "No, sir." *Not if my life depended on it.* After that, I stayed focused on my work.

By Friday night, I realized there was a lot about Zach that I didn't know—and probably couldn't even guess. But I *did* know three very important things: 1) I knew that no one at school was going to laugh at Miyoko again—or me either probably—because there was a rumor that Miyoko could kill a person just as fast as she could look at them; 2) I knew four out of the five recipes I was going to send to *Southern Living* on Monday; and 3) I knew another reason why I didn't want my mom to marry Keene, on top of all the other reasons. If Mom married Keene, then she'd be starting fresh on her dream of having a family. If you ask me, that's an awful lot like starting fresh on dinner. And if Mom was starting fresh, then that made me a kind of leftover, didn't it? Yes, I was a leftover from her previous attempt at marriage and family.

Here's the thing about leftovers: Nobody is ever excited about them; they're just something you have to deal with, like Keene has to deal with me. No matter how hard you try, leftovers are never exactly what they used to be—and I'm not either. If you ignore them or forget about them, they start to stink, and if you try to serve them alongside a freshly made meal, they never fit in quite right—do you want leftover spaghetti with your fajitas? Ugh! Leftover spaghetti is the worst! See, when you reheat spaghetti noodles, they overcook and turn to mush. And no matter what you do with them, leftover spaghetti noodles stick together in clumps. They get hard in some spots and soggy in others. If you want my opinion, it's best to just throw leftover spaghetti away. And I was leftover spaghetti! No, I was *worse* than leftover spaghetti, and a lot more trouble—does a visit from the fire department ring any bells?—and I *couldn't* be thrown away.

I'd wanted to talk to Aunt Liz about all of this as it was taking shape in my mind that afternoon, but I didn't think I could do it yet without crying. Somewhere between the cooking and the homework, Parents' Night, and all the upset over Keene and my purple cake, I'd gotten tired. *Too* tired. And when I get too tired, I get sort of wilty and weepy and turn to mush. Like leftover spaghetti. Yuck.

So, there I was, standing in my kitchen at 9:08 on Friday night with tired-tears in my eyes, trying to decide whether to go to bed or try out Great-Grandma Russo's recipe for lasagna. I really was tired. And we didn't have the exact ingredients the recipe called for. And Mom and I had already eaten dinner. And I already had four recipes to send to *Southern Living*, and really, four was enough, wasn't it? I decided to go to bed.

But before I reached the bottom step, I heard Keene's voice somewhere in the back of my mind: *Cecily, you don't really believe she can win the contest, do you?* Then I heard my mother's voice: *Of course I believe she can win.* Two things carried me back to the kitchen that night, when I really wanted to go to bed: 1) I really wanted to prove Keene wrong; and 2) I really wanted to prove Mom right.

I filled our biggest pot with water, added a dash of salt, and set the pot on the stove to boil. Then I began preparing all my ingredients in bowls—*mise en place* style. Since we didn't have some of the ingredients, I had to get creative and come up with substitutions, but I like being creative. Soon I didn't feel tired anymore. I was having fun.

I was having so much fun that I became television star Fizzy Russo, of *Fabulous Foods and Feasts with Fizzy Russo*. I smiled for the cameras and pretended my mismatched bowls matched. For a few seconds, I was tempted to put on my mom's engagement ring for when the cameras zoomed in on my hands. I mean, the ring was *right there* on the windowsill above the kitchen sink—where she leaves it whenever she cooks or cleans. But I figured if I could pretend glass bowls, then I could pretend rings on my fingers, too, and I left Mom's ring where it was.

"Now, if you don't have Italian sausage," I told my pretend audience, "then you can use any kind of sausage or hamburger if you like." I smiled sweetly.

"Fizzy?" Mom said, coming up behind me.

I jumped, let out a spastic *aaaah!* and then turned.

"Fizzy, what are you doing? It's almost ten o'clock," Mom said, pulling her cardigan closed tight over her pajamas.

"I'm making lasagna," I said, as if it were the most natural thing in the world.

Mom gave me a serious look of disapproval and I thought it was a good thing that I wasn't wearing her engagement ring.

"It's for the contest," I said. "We can eat it all weekend . . . I bet Keene will like it, too." I thought that was a nice touch.

And it worked. "All right," Mom said, softening, "but straight to bed as soon as you're finished. We have a lot of shopping to do tomorrow."

"Yes, ma'am," I said as Zach's words echoed through my mind: *It's best for everybody if you just say whatever the adults want to hear.*

Chapter 14

Mom stood in the three-way department store mirror wearing a long, shimmery, whipped-cream-colored gown.

Tired from the late-night lasagna and this morning's parade of bridal gowns, I sat on the floor cross-legged in front of her.

"What do you think, Fizzy?" Mom asked.

At that moment, I thought she was the most beautiful woman I'd ever seen. And it gave me hope—because Mom has the same kind of hair and eyes as I do—she even has a few freckles sprinkled across her nose, like sugar on a cookie.

"Fizzy?" Mom said, uncertain. "What's the matter? You don't like it, do you?"

"No, ma'am," I said. "I *love* it. It's beautiful. You're beautiful."

Tears of happiness glittered in Mom's green eyes and I knew then that I could never tell her about feeling like leftover spaghetti. Never, because I loved her and I really did want her to be happy. I really did want her dreams to come true. I really did.

Mom took one last long look in the mirror and then said, "All right, it's your turn."

I tried on two itchy peach-colored dresses that Mom—naturally—loved.

"What don't you like about them?" she asked, tilting her head to one side as she looked at me.

"This stuff," I said, lifting the top layer of the skirt to show her the stiff, netted, cheese-cloth-like material underneath.

"Those are crinolines," Mom said.

I raked my fingernails up and down my thighs.

"Well," Mom said, "we can't have you scratching yourself like that at the wedding, so I'll go look for something else."

The door to the dressing room had barely clicked closed before I had that dress up over my head and off. Then I just stood there awkwardly in my underwear, waiting for Mom to bring more dresses.

A few minutes later, I heard Mom coming down the hall, saying, "I've found it, Fizzy. This is the one. I just know it."

I cracked the door and Mom handed the dress through. "But, Mom," I said, "it's purple."

"To match your cake," she said.

I loved the dress. Not only did it not have itchy crinolines, it was the most elegant shade of purple—kind of silvery.

After we'd paid for our dresses, Mom and I went looking for matching shoes.

"What do you think of these?" Mom asked, holding up a shoe.

I searched Mom's face, trying to decide if she was joking. "Ummmm . . . those are *gym* shoes," I informed her.

"I know," she said. "When I saw Coach Bryant at Parents' Night, he suggested that you bring an extra pair of gym shoes to keep at school."

"Okay," I said too quickly, hoping that would be the end of it, *really* hoping that Coach Bryant hadn't said anything else—like how I scarcely participate in gym class, how bad I am at it when I do, and how I'm always last-picked for any team—these are not topics I'd like to discuss. At all.

But apparently, Coach Bryant hadn't said anything else, because Mom moved right on to a delicate pair of pearl-beaded, whipped-cream-colored high heels.

"This is the best lasagna I've ever tasted," Keene said that night at dinner.

"Fizzy made it," Mom said proudly.

"Unbelievable," Keene said.

I thought it was pretty good myself. In fact, I thought it was *Southern Living* Cook-Off good and I was glad I'd taken the time to jot down the substitutions I'd used.

As Mom and I cleared the dishes, Keene said, "So are you two ever going to show me those dresses?"

"Not mine," Mom said right away. "It's bad luck for the groom to see the bride's dress before the wedding. But you can see Fizzy's."

"Okay," Keene said agreeably.

I rolled my eyes. He was just being nice for Mom's sake. I knew he didn't really want to see my dress. Who cares what you put on leftovers? You can dress them up and garnish them all you want, but they're still leftovers and everybody knows it.

Of course, Mom brought my new dress down to the dining room and showed it to Keene anyway.

"I like it," Keene said, nodding. "I like it a lot."

Mom smiled at me, as if to say, *See? See how nice Keene is?*

I didn't respond.

"I'll just run upstairs and put this back," Mom said, giving Keene some sort of look as she pulled the plastic back down over my dress.

I was still trying to figure out what the look meant when Keene said, "Have a seat, Fizzy."

Oh, I thought. *It was a talk-to-her look.* I sighed and dropped into a chair.

"I really like your dress," Keene said. "I like the color purple."

"Thanks," I muttered.

He folded his hands into a steeple on top of the table.

I waited for him to speak, thinking of the nursery rhyme I'd learned at church when I was little: *Here is the church, here is the steeple . . . open it up and see all the people.*

"You know, Fizzy," Keene said, "I don't have any kids. I don't really even *know* any kids."

I just stared at him, not knowing what I was supposed to say to that. *I know? Yes, sir, it's pretty obvious? I'm sorry? Good for you?*

Keene took a sip of iced tea and set his glass back down on the table. "And I've been by myself for a long time."

I watched the beads of condensation slide down his glass.

"And when you're by yourself, you can be selfish because you only have yourself to think about. Do you understand?"

I nodded absently, without taking my eyes off his tea.

Keene shifted uncomfortably. "What I'm trying to say is that I'm still learning."

I nodded again.

He reached out like he was going to touch my hand, but I pulled it back—so that he couldn't.

Keene sighed. "Look, Fizzy, I could tell you that I love you and that I'm excited about becoming your stepfather, but the truth is, you and I don't know each other well enough for any of that."

"I know," I mumbled into my lap. "I don't expect you to love me."

Keene held up his index finger and said, "But I do. I expect to love you, in time, just like I grew to love your mother in time. It's just going to take some time, that's all."

What is it with adults and time? An adult's solution for almost any problem is time (well, except for Coach Bryant, whose solution is fresh air and exercise). *In time, you'll understand. You'll get used to it in time. You'll make friends in time. In time, it'll all be okay. In time, maybe I'll love you. Maybe. Maybe not.*

"Don't worry about it," I said, feeling the tears coming and not knowing why. I stood.

"Fizzy, please, listen to me," Keene said.

I sat back down.

"I have every reason to believe that we can all be very happy together. If I didn't believe that, I wouldn't have asked your mother to marry me . . . and she wouldn't have said yes."

I wanted to believe him. I really did.

"And we still have time. I promise you, Fizzy, if I ever start to think we can't be happy together, all of us, then your mom and I won't get married—at least, not for a while."

"Okay," I said, closing my eyes—so the tears couldn't escape—and drawing a ragged breath.

"For now, let's try to get to know each other," Keene said. "Let's try to become friends, okay?"

I opened my eyes. "You want to be my friend?" I said in disbelief.

"Yes, I do. I can be a good friend, Fizzy. I'm a really good friend."

Not to be outdone, I said, "I'm a good friend, too."
Keene smiled.
I smiled back.

That night, I added to my Things I Like About Keene List:

3) *Likes my lasagna.*
4) *Likes the color purple.*

I like those things, too. And after all, I'd started friendships over lesser things. Recently, I'd started one friendship over a marble, and another over a common route to school.

Plus, I had to give Keene some credit for not hating me the way I hated leftovers. Okay, so he didn't love me, but he didn't hate me either.

Chapter 15

Standing outside the massive double front doors at Miyoko's house before my first sleepover in Lush Valley, I felt very small—and nervous—about everything, especially where I lived. I hoped Mr. and Mrs. Hoshi wouldn't ask which big Lush Valley house was mine, because somehow I knew the answer would disappoint them.

Naturally, as soon as Mrs. Hoshi opened the door, she said, "Hello, Fizzy! We're so glad you could come. Did you have to come far? Where do you live?"

Mrs. Hoshi was petite and pretty, like Miyoko, and very well dressed—except for the house shoes on her feet.

I turned and waved at Mom—who'd stayed in the car—and then stepped onto the fancy marble floor in the foyer. "Actually, my family has two houses," I heard myself say.

"Oh, how nice!" Mrs. Hoshi said. "Where are they?"

"Um . . . well, my mom has a town house here in the valley," I said as Miyoko took my suitcase and set it at the bottom of the staircase.

"I know just where that is," Mrs. Hoshi said, helping me out of my coat.

Of course she did. They were the only town house complexes—the only *small* homes—in all of Lush Valley. "And my dad has a house outside the valley, on Candlelight Way."

"Shoes go there," Mrs. Hoshi said, pointing with a manicured hand to a black rubber traylike thing hidden under an ornately carved bench.

I didn't really want to take my shoes off because I hadn't expected to and therefore hadn't inspected my socks. But I did it anyway. My socks seemed okay: no holes and not too dingy.

Mrs. Hoshi hung my jacket in the coat closet and said, "Come on in the kitchen and I'll make you some spiced tea."

Miyoko and I followed Mrs. Hoshi into the chef-quality, state-of-the-art kitchen, where she fixed each of us a cup of hot tea.

"So your parents don't live together?" Mrs. Hoshi said, looking concerned, as she handed me a steaming mug.

"No, ma'am," I said. "But everything's okay."

"Are they married?"

"Um . . . not to each other," I said. Then I took a sip of tea, which tasted a lot like mulled apple cider, only not sweet.

Mrs. Hoshi looked at me like, *Then how could everything be okay?*

Exactly, I thought, but what I said was, "They still like each other and everything—they're still friends." *It's best for everyone if you just say whatever the adults want to hear.*

Mrs. Hoshi pursed her lips. I figured either she didn't believe me or she didn't think friendship was enough. Since I didn't know which, I didn't know what to say next.

Luckily, Miyoko did: "Mom, I'm going to take Fizzy upstairs and show her my room now."

"Leave your tea here," Mrs. Hoshi instructed.

Fine by me, I thought, because what good is tea without sugar?

It turned out that Miyoko didn't have a "room." What she did have was a "suite," complete with sitting area, study area, bedroom, and full bathroom.

"Wow," I said. "This is . . . *really* nice."

"Thanks," Miyoko said, quietly closing the door behind us. "Listen, I'm sorry about my nosy tiger mom."

I shook my head. "That's okay. And anyway, I didn't think your mom was *vicious* or *tiger*like."

Miyoko laughed. "'Tiger mom' is an expression for a very controlling mom who demands perfection in all things."

"Does yours demand perfection . . . in all things?"

"Pretty close."

I plopped down on the bed. "Then it's a good thing I don't live here."

"Yeah," Miyoko said, perching on the edge of the bed beside me. "It's not much fun."

"I'm sorry," I said.

Miyoko nodded. "I'm sorry about your parents, too."

I started to say no, to defend my parents. But then I realized I wasn't having much fun with them either lately. Instead, I just said, "Thanks."

There was an awkward silence.

I broke it by confessing, "Um, I just have a regular room." I wanted to get that straight right up front, because it was the truth.

"That's fine—that's great—I'm sure your room is great," Miyoko said.

Since I was already confessing, I continued, "And I don't know why I said that—about my family having two houses—it was obnoxious."

Miyoko shrugged. "My dad says that when people start talking about the things they have, or the things they're going to get, it means they're scared."

I thought about this. "He's right," I decided out loud. "I felt scared."

Miyoko smiled. "That's okay. There are a lot of scared people in Lush Valley."

I knew what she meant: There were a lot of obnoxious—scared—people in the valley. I didn't want to be one of them. *Note to self*: *When you're scared, don't talk.*

At dinner, which seemed a little strange and a lot healthy—tofu, brown rice, and plain steamed vegetables—Mrs. Hoshi placed an egg timer on the table and set it for thirty minutes.

"Is your dessert still in the oven?" I asked hopefully.

"No," Mrs. Hoshi said. "We don't eat sugar."

"Oh," I said. "Well, that's . . . good." I guessed it was good but I couldn't help feeling a little sorry for the Hoshis. They probably didn't have any butter either. What's a life without sugar? And butter?

"I'm a slow eater," Miyoko explained. "Dinner is thirty minutes. When the timer goes off, it's over."

"Oh, sure, okay," I said, like timing a family dinner was the most natural thing in the world, even though I'd never heard of such a thing.

"It's important not to waste time," Mrs. Hoshi said, "because time is what your life is made of, Fizzy."

"Yes, ma'am," I said.

Mrs. Hoshi lowered her eyes, to the little pile of broccoli on my plate—which I had planned to hide under some tofu. "There are children starving to death in Africa," she informed me.

Please, feel free to ship my share of broccoli to them . . . for the rest of my life, I thought, but I said only, "Yes, ma'am." And then I ate some broccoli. It was pretty bland—it was *all* pretty bland—I don't think the Hoshis have salt either.

After dinner, Miyoko took me upstairs, where she apologized for both of her parents.

"It's okay," I said easily. "They're parents."

Miyoko just stared at me.

"All parents say dumb stuff," I explained. "My mom says stuff like, 'Close your mouth and eat your dinner.'"

Miyoko giggled. "Yeah, I fail to see how eating my broccoli helps starving children."

"You're probably just too young to understand. Give it time." I grinned.

Miyoko grinned back. "In the meantime, well . . . you're only young once!"

"Thank heaven," I said, "because it's harder than it looks."

"Just be yourself," Miyoko said.

I rolled my eyes. "Do I have any other options?"

We laughed until we could barely breathe.

After that, Miyoko taught me some pretend-karate—some chop-choppy things and some kick-kicky things. "Have you seen *The Karate Kid* movie?" she asked.

"Yeah, a long time ago."

"Okay, here's how you do the big kick at the end," Miyoko said, lifting her arms high while effortlessly balancing on one foot.

I tried to mirror her, raising my arms and one leg—I was a little shaky.

"Are you ready?" she asked.

"Yes," I said confidently. And then I promptly fell over on the carpet. *WHUMP!*

"Oh! Are you okay?" Miyoko crouched beside me.

I couldn't answer her because I was laughing so hard. Miyoko laughed, too.

"What's going on up there?" Mrs. Hoshi hollered.

"Nothing!" Miyoko hollered back. Then she lowered her voice and said, "We should probably save that one for another day . . . but we could watch the movie—study the moves."

I propped myself up on my elbows and nodded.

We started *The Karate Kid*. But we didn't finish it because Mrs. Hoshi came upstairs and announced that it was bedtime.

"It's not even ten o'clock," I whispered as I watched Miyoko floss her teeth in the bathroom.

She removed her fingers from her mouth. "I know, but my dad says it's better for your body to stay on the same schedule all the time—he's a doctor."

"Oh," I said. "My dad's a dentist—he'll be thrilled to know you actually floss."

Miyoko smiled. "Sorry we have to go to bed."

"No, it's okay—really," I said. "I've had a great time at my first sleepover in the valley."

"Me too," Miyoko said. "I've had a great time at my first sleepover. Ever."

"*Ever?*" I repeated.

"Ever. My mom says school isn't for making friends; it's for learning."

I shook my head sadly. "And here you've gone and made a friend when you were supposed to be learning," I teased.

Miyoko smiled. "I know. I've gone wild."

"Miyoko?" I whispered into the darkness, not sure if she was still awake.

"Hmmm?"

"Do you know Zach Mabry?"

Miyoko rolled onto her back and whispered to the ceiling, "The guy Buffy Lawson likes?"

I sighed. "Yeah."

"Yeah—he's cute."

"Yeah," I agreed.

Miyoko seemed to wait for me to say more.

"Do you think Zach will invite Buffy to the Valentine's dance at school?" I asked, because everybody knew about Buffy's crush on Zach. Plus, Buffy goes out of her way to be all... *pretty* right in front of Zach's eyes. I happen to know that after math, Buffy's next class is down the hall on the right, but she always walks left out of math because Zach does—then she has to race back before the second bell.

"I don't know," Miyoko said. "Buffy's so... Buffy."

"Yeah," I said, "and I'm so... *not* Buffy."

"I'm glad," Miyoko said.

I smiled. "G'night, Miyoko."

"Night," she said, and then she rolled back onto her side.

I felt a little homesick lying in Miyoko's bed, but it didn't bother me—because you're *supposed* to feel homesick when you're away from home.

Chapter 16

On Friday night, when I hauled my suitcase and myself downstairs, ready to leave for the Valentine's dance at school, Mom demanded, "Are you wearing lipstick?"

"No, ma'am," I said, and it was true. I wasn't wearing lipstick; I was wearing strawberry Jell-O. (With just a little water and a packet of strawberry Jell-O powder, I'd created a natural-looking lip stain—at least, *I* thought it looked natural.)

Mom gave me a doubtful look as I put on my coat, but Keene was there waiting, so she didn't say anything else.

"All set?" Keene asked.

Mom and I nodded and followed him out the door. Mom and Keene were dropping me off at school on their way to see a movie; Dad would pick me up from the dance, since it was his weekend.

The car ride was silent. And tense. I began to regret offering Miyoko a ride to the dance—she would surely feel that something was wrong as soon as she got in the car—but the thought of walking into the dance alone . . . with a suitcase . . . had completely clouded my judgment.

As soon as we turned into the driveway, Miyoko appeared, hurrying through the cold night air toward our car. Mrs. Hoshi waved, from just inside the front door—instead of coming out to inspect us, the car, and its safety features, which I thought was nice of her. It turned out that Aunt Liz had decorated the Hoshis' house. After Mrs. Hoshi had learned that I'm Liz Talbott's niece, she was a lot nicer.

We all waved back and even that small, unified gesture felt weird to me—like we were trying too hard to look like a normal family when we knew darn well that we weren't.

When Miyoko pulled the car door shut, I said, "Miyoko, this is my mom and her . . . um . . . Keene."

"It's so nice to meet you," Miyoko said. "Thank you for giving me a ride to the dance."

"Our pleasure," Mom said. "It certainly is cold outside, isn't it?"

I knew then that Mom was as uncomfortable as I was, because she only resorts to talking about the weather when she gets really nervous.

"Yes, ma'am," Miyoko said.

"But spring will be here before we know it," Mom said. "In six weeks, Lush Valley will be lush and green again, with blue skies and sunshine."

"And wedding bells," said Keene.

Mom reached over and sort of petted Keene on the back of his neck while he drove. "We're getting married in April," she explained to Miyoko.

Miyoko gave me an uncertain look.

My stomach felt icky. I took a deep breath and held it.

Miyoko reached over and gave my hand a sympathetic little squeeze.

We could hear the music pounding outside the school before we even opened the door.

Zach was sitting on the stairs in the foyer when Miyoko and I entered. He stood and gave us a lopsided grin when he saw us.

Miyoko smiled back and gave me a little shoulder bump.

Zach trotted down the stairs and came to a stop in front of us.

"Zach, this is my friend Miyoko," I said. "Miyoko, this is my friend Zach."

"Hey," Zach said.

"Hi," Miyoko said.

"This way, ladies," he said, extending his arm toward the gym.

A HAPPY VALENTINE'S DAY banner hung over the door leading to the gym—even though Valentine's Day wasn't until tomorrow. The gym was dark except for little spots of light thrown out by two disco balls hanging from each basketball hoop. Because the gym also serves as an auditorium, there was a stage up front, cluttered with coats and purses piled one on top of another. Miyoko and I went to the stage and took off our coats. I pushed some others aside, hefted my suitcase onto the stage, and did my best to cover it with my coat. Then I looked around to see if anybody had noticed, but no one was paying attention.

Kids danced in little groups—of mostly girls. A lot of boys stood around, leaning against the walls. A few boys chased each other around the perimeter of the gym.

Zach tilted his head as if to say, *C'mon.* Miyoko and I followed him into the crowd, where we found a spot and began dancing together. The full skirt on Miyoko's red dress swished in a pretty way as she moved and made me rethink wearing my regular old jeans. But I preferred jeans over my church dresses, and I'd worn my favorite pale-pink thermal shirt, which I had reasoned was sorta Valentinesy. I remembered my Jell-O lips then and they boosted my confidence just a little.

When a slow song came on, I started to move off the floor, but Zach grabbed my hand.

I leaned into his ear—to be heard above the music—and said, "I don't know how to slow dance."

"I'll show you—it's easy," Zach said, his breath warm on my ear. Zach said something in Miyoko's ear, too, and she nodded. He stepped in front of her, giving Miyoko his back, as she placed her hands on his shoulders. Zach placed my hands on top of Miyoko's, on his shoulders, and then rested his hands—light as feathers—at my waist. I fought the urge to laugh—I'm very ticklish—plus, I felt nervous, which sometimes causes me to laugh inappropriately. Then we all three rocked from side to side in time with the music while rotating in a circle. My eyes went from Miyoko to Zach to my feet—to make sure they didn't step on Zach's feet—and back again.

Miyoko looked all around the gym, which I understood—it was weird if we stared at each other.

Zach kept on smiling, seemingly to himself more than anybody else, which finally caused me to lean forward and say into his ear, "What?"

The smile widened. "Man, I must look really cool right now—these other guys can't work up the nerve to ask even one girl to dance, and here I am dancing with two!"

I did laugh then.

But Buffy and Christine must've agreed with Zach, because when the next song came on—a fast one—they pushed their way into our little dancing threesome, on either side of Zach.

Zach took a step forward and said loudly, "I'm thirsty. Fizzy, Miyoko, y'all thirsty?"

Miyoko and I nodded and followed Zach out of the gym, leaving Buffy and Christine to dance with each other.

We filled paper cups with Valentine's punch from the table in the hallway, between the bathroom and the cafeteria, and we stood off to one side of the hall drinking it. I'm pretty sure the punch was just cherry Kool-Aid—I bet I could make lip stain with that, too.

"Thanks for teaching me to slow dance," I said.

"Thanks for making me the coolest guy in the building." Zach grinned.

Miyoko downed her punch and announced that she had to go to the bathroom.

"I'll come with you," I said.

"Think I'll wait here," Zach said, grinning again.

Miyoko stepped in front of the bathroom mirror and leaned

over the sink to inspect a pimple on her nose. "It's grown since I left the house," she reported. "And it has an eyeball! A big white eyeball! Ugh!" She touched the blemish. "And it's throbbing! I think it has a heartbeat!" She turned from the mirror. "Fizzy, I think I'm growing another person on my nose!"

I laughed. "It's not so bad." The truth is that it was pretty bad now that I was looking at it, but I hadn't really noticed the pimple until Miyoko pointed it out.

"I'm going to call him Ogle," Miyoko announced. "Ogle, the nose pimple-person." With that, Miyoko went into a stall and closed the door.

I was just about to unlock my stall and come out when Buffy and Christine came into the bathroom talking. "Did you see Fizzy's lipstick?" Buffy said, in a way that let me know she wasn't a fan of my Jell-O lips.

"How could I not?" Christine said. "People across the street can see it from here—and I'm sure they think it's tacky, too!"

I felt my face heat up. I took a deep breath, unlocked the door, and forced myself to walk calmly to the sink.

Buffy was reapplying lip gloss in the mirror while Christine fluffed her hair. When they saw me, they exchanged a smirky look.

I'd just turned off the faucet and was reaching for a paper towel when I heard the lock open on Miyoko's stall door. I looked up at the reflection of her door in the mirror. It remained closed.

Then, *BAM!* Miyoko kicked the door open and burst out of the stall into a lunge, her knees bent and her hands poised to do the chop-choppy thing in the air. Buffy, Christine, and I all jumped.

"Say something else about my friend Fizzy," Miyoko dared them. "Go ahead. I *want* you to."

"Sorry," Buffy mumbled as she scurried from the bathroom.

"Yeah," Christine murmured, following close behind her.

Note to self: It's never smart to mess with a girl who has a pimple with an eyeball—or her friends.

When I was sure Buffy and Christine were gone, I turned from the mirror to face Miyoko. "That was *awesome*!" I gushed. "How did you do it?! If *I* kicked a door open, it would just bounce back, hit me in the face, and probably knock me out!"

"I didn't consider that," was all Miyoko said, and then she washed her hands.

I looked at my Jell-O lips in the mirror and wondered if they were "tacky."

"You look great," Miyoko said, reading my mind. "You, me, Ogle—we all look great. Buffy's just mad because Zach's not dancing with her."

Zach was still waiting for us in the hallway. When he smiled at us, Miyoko said, "Are you looking at my pimple?"

Zach took a step backward, showed Miyoko his palms, and said, "Not me." Then he showed her his teeth.

Apparently Zach's I'm-not-a-violent-maniac smile also doubled as his I'm-not-looking-at-your-giant-zit smile, too.

"I've named him Ogle," Miyoko announced, "because even if you're not watching him, he's watching you—he's like the *Mona Lisa* of pimples." She looked around self-consciously.

"Well, nobody else is looking either," Zach said, letting his hands fall to his sides. "We're all too worried about our own Ogles to notice yours."

I thought about my suitcase then and realized that Zach was right. As we all walked back to the dance, I couldn't help wondering what Zach's Ogle was, but I decided it'd be rude to ask.

Later, when my dad arrived a few minutes early and I said good-bye to Miyoko, Zack seemed surprised. "Thought y'all were having a sleepover," he said.

It took me a minute to realize that Zach had thought that because of my suitcase, and it made me feel much better. *Note to self: Suitcases don't have to say, "My family is a big, broken mess and so am I!" They can also say, "I am a totally normal person who has friends, and I'm sleeping over with one of them! Yay!"*

Suzanne was sitting on the couch eating a box of Valentine's chocolates when we walked into the house.

"We have a Valentine's surprise for you," she said in between bites of chocolate. She smiled at Dad.

He smiled back.

I smiled, too, thinking, *I could go for some chocolate.*

"We're having a baby," Dad just blurted out, the same way he might've said, *We're having blueberry pancakes for breakfast.*

I felt my smile slip, but I hoisted it back up and said, "That's great."

Suzanne smiled and nodded. "Of course, it'll be some

months before we actually get to meet the baby." She patted her stomach.

"That's great," I said again, staring at Suzanne's stomach, which I now noticed stuck out *almost* as far as her chest—how had I missed this? That baby must've been doing some serious growing lately!

Dad stared at me.

"Wow, um . . . well, I'm pretty tired," I said, "and I still have to unpack, so . . . um . . . good night." I waited a few more seconds—in case someone wanted to give me Valentine's candy—but no one did.

That night, I lay awake in bed wondering what it would be like to have a baby in the house. *It'll probably be like having a puppy,* I told myself. *Okay, a bald-ish puppy, but still. Who doesn't want a puppy?*

Chapter 17

At home, things were as awkward as ever between Keene and me—even after our "talk." It felt as if we were both waiting for the other to take the first step in a dance neither one of us knew how to do—it was a lot harder and more complicated than The Middle School Slow Dance, that's for sure. Eventually, we gave up waiting for each other and instead waited for the song to end. I waited for Keene to go home. And he seemed to wait for me to go anywhere: to Dad's, to Miyoko's, or even just back to my room. But I knew it was only a matter of time before the song—and the dance—no longer had an end.

Of course, I ignored that thought—Mom's wedding—as much as possible. When I couldn't ignore it—due to wedding invitations, seating arrangements, fittings, gift registries, cake tastings, and bridal showers—I hoped for something, *anything*, that would stop the wedding. A major fight between Mom and Keene, a natural disaster, an outbreak of bubonic plague—any of that would've been fine with me.

Well . . . except that a disaster or disease would've kept me home from school—and school had become one of the few places I could escape to where I pretty much knew what to expect and what was expected of me and where I'd actually begun to feel comfortable. Mostly when I was with Zach and Miyoko. The three of us did everything together: We ate lunch together, sat together during assemblies, stayed together during field trips, and walked home together whenever Zach didn't have to stay after school.

All of these things made the very, very short list of Things I'm Happy About.

Buffy Lawson, on the other hand, seemed less than happy with our lunchtime seating arrangement—she stared at us and did a lot of eye rolling. Obviously, she couldn't stand me.

So I was pretty surprised when she shouted and waved as Zach, Miyoko, and I carried our lunch trays through the cafeteria that Thursday.

"Fizzy! Fizzy, come sit with us!" Buffy called out.

Now, I'd heard my name, but still I stopped and looked around, thinking there must be some other Fizzy with whom Buffy Lawson was willing to be seen. No other Fizzys in sight.

Miyoko stood frozen beside me, her eyes fixed on Buffy like she was a dangerous and deadly spider.

"Um, I'll just go see what she wants," I said.

"Want me to come with you?" Zach asked.

Honestly, I wanted Zach *and* Miyoko to come with me, but there wasn't enough room for all of us, and I didn't want Miyoko to be left alone. "That's okay," I said.

Miyoko opened her mouth but then closed it again. She nodded without looking me in the eye.

When I started toward them, Buffy and Christine immediately made room for me, right next to Buffy. *Wow*, I thought as I sat down. I felt very important.

"So," Buffy said, tossing her hair over her shoulder, "what's going on?"

"Um, not much," I said. "What's going on with you?"

Buffy and her followers exchanged amused looks.

Finally, Buffy said, "I thought maybe you'd like to give us the scoop on Zach Mabry," like she was doing me a big favor.

I squirmed a little in my seat, not sure what to say. Eventually, I came up with, "He's nice."

Buffy laughed, and then all her friends laughed, too. "I know *that*," she said, still giggling, "but does he have a girlfriend?"

"Oh. I . . . um . . . I don't know," I said, realizing I'd never asked Zach that question. I felt a little disappointed at the thought that he might have a girlfriend.

Buffy looked disappointed, too, but she recovered quickly. She recovered and ignored me for the rest of lunch.

Her friends ignored me, too, so it was a really long lunch—not unlike dinner at home. I wanted to get up and go sit with Zach and Miyoko, but since we're not allowed to change seats after we sit down in the cafeteria, I just sat there.

Mostly, Buffy and her followers talked about their clothes, shoes, and bags. I remembered how Miyoko's dad had told her that people who talk about their stuff are scared. I wondered if they were scared. When they moved on to discuss other people's clothes, shoes, and bags, I *knew* they were scared. They were right to be scared—of each other—because they all could be really mean. I was glad when Buffy gave the nod indicating that it was time for their "big catwalk moment" and that lunch was over.

While I was dumping my tray, Mrs. Hunt, the lunch lady, crooked her finger at me, as if to say, *Come here.*

I went to her. "Ma'am?"

Mrs. Hunt smiled and came out from behind the stainless steel buffet line. "You know, a true friend is hard to find."

"Yes, ma'am." I nodded.

"And you've found *two*," Mrs. Hunt said, still smiling.

I knew she was talking about Zach and Miyoko. "Yes, ma'am . . . I . . . I'm pretty lucky."

"You are," she said. "I used to worry about you, but . . . you're going to be just fine, Fizzy." Then she fished a piece of foil-wrapped chocolate out of her pocket and handed it to me.

"Thank you, Mrs. Hunt," I said.

Lunch ladies are a lot like office ladies, I thought as I got in line to leave the cafeteria. *They're quiet, but they know* everything.

"What did Buffy want?" Zach asked, using one hand to shield his eyes from the sun as we walked home that afternoon.

Miyoko kept walking and pretended she wasn't interested, but I knew she was.

"She wanted to know if you have a girlfriend," I said.

Zach rolled his eyes. "And?"

"I said I didn't know."

"Anything else?" Zach said.

"Yeah, I gave her a really juicy tidbit about you."

Zach looked at me expectantly.

"I told her you're . . . *nice*," I said.

Zach laughed, and then Miyoko and I did, too.

"So," Miyoko said hesitantly, "did Buffy ask you to eat lunch with her again tomorrow?"

I looked at her. "No."

Miyoko kept her eyes on the sidewalk and said only, "Oh."

I gave her a little bump with my shoulder. "And I wouldn't eat with Buffy again even if she *begged* me."

Miyoko smiled and gave me a little bump back.

"She's boring, right? See? I told you," Zach said.

"'Boring' isn't the first word that comes to mind," I told him.

"What is?" Miyoko wanted to know.

I thought about it. "Scared. And mean."

"Oh, man, that *is* a bad combination," Zach said.

Chapter 18

Sadly, Mom and Keene's April wedding day arrived without incident. Since I'd been allowed to invite one guest to the wedding, Miyoko and I sat together that Saturday morning in the haze of perfume and hairspray that filled Mom's "dressing room"—which was really a Sunday school room at church. As I watched everybody from makeup artists and hairdressers to friends fuss over Mom, I asked, "Do any of y'all know how to cover up freckles?"

"You don't like your freckles?" Miyoko said.

I turned and gave her a look that told her that was just about the dumbest question I'd ever heard. "How would you like it if you had *these* all over *your* face?" I asked, pointing.

"I'd love it," Miyoko said, "but I'll help you hide them if you want me to."

"You will? . . . Right now?"

Miyoko nodded and eyed the makeup artist's kit sitting on the table next to us.

I felt like a cake being frosted, but when Miyoko finished slathering makeup on my face, I couldn't see a single freckle! Okay, so my face wasn't the same color as the rest of me, but the important thing was that my freckles had completely disappeared.

"This is great," I said, looking in the mirror attached to the lid of the makeup kit. *Note to self: Makeup is a mini-miracle. Get some.*

Mom didn't really get a good look at me until her bossy friend Eulalie walked her down to wait outside the doors of the church sanctuary, where I was already standing. But when she did, Mom gasped and her eyes bulged—it was a good thing Eulalie was there because she kept Mom from walking face-first into a column.

I smiled brightly and nodded at Mom, as if to say, *I know! It's shocking how much better I look without the freckles, isn't it?*

Just as Mom opened her mouth to say something to me, Eulalie grabbed my elbow and pulled me toward the sanctuary, speaking urgently. "It's time. Now remember: Step together, step together, step together—*slowly*." Then she shoved me through the doorway and down the aisle.

I stood in my silvery-purple satin dress next to Mom while she and Keene said their vows to each other. When they finished, the preacher said, "You may kiss the bride." So Keene did. He kissed Mom right there in church, in front of everybody—and we're not talking about a little kiss; we're talking about a big, long, disgusting, spit-swapping kind of kiss. I couldn't believe it! Yuck!

Just when I thought things couldn't possibly get any worse, the preacher said, "Ladies and gentlemen, it is my great honor to introduce to you, for the very first time, Mr. and Mrs. Adams."

Mrs. Adams? Who was Mrs. Adams? That's when I realized my mom was Mrs. Adams. We didn't even have the same last name anymore! I mean, I knew that wives usually took their husbands' last names, but I hadn't exactly done the math here—I *hate* math.

My nose and eyes stung as I watched Mr. and Mrs. Adams walk hand in hand up the aisle and out of the sanctuary. I felt like a stranger to them, somebody on the outside looking in—a pitiful face pressed against the glass—instead of part of their family. But I hadn't really expected to be part of the Adams family, had I?

At least I wasn't the only one who was no longer Mom's family. At least I had Dad, Aunt Liz, Uncle Preston, and the entire Russo family to keep me company on the hacked-off branch of the family tree.

The preacher touched my arm lightly and motioned for me to follow Mom and Keene.

Miyoko was waiting for me by the double doors to Fellowship Hall, which is the fancy name for the cafeteria in the basement of our church. I could hear a piano being played and the hum of people talking all at once inside, where Mom and Keene were having their wedding luncheon.

"Hey," Miyoko said, smiling. "You looked great up there—not a freckle in sight."

I wanted to tell her that she looked great, too. Miyoko wore a deep-eggplant-colored dress that showed off her beautiful, freckle-free skin and dark shiny hair, which was pulled up into a perfect ballerina bun. But I felt too tired to talk. Suddenly, all I wanted was to go home, climb into bed, and pull the covers up over my head. But I couldn't.

Miyoko followed me through the buffet line to the kids' table. Since we were the only kids there, it was just the two of us. *I don't belong,* I thought to myself again as we sat down.

Miyoko bit into a cheddar cheese cube and watched me.

I stared at the rubbery-looking chicken on my plate. I wasn't hungry.

"What's wrong, Fizzy?" Miyoko finally asked.

I didn't even know where to begin, so I shook my head and said, "I just want normal parents, you know?"

Miyoko only blinked at me.

So I explained, "You know, *normal parents*, who have the same last name as me and don't kiss in public—in church! It's just so weird. They're so weird. It's embarrassing."

Miyoko nodded, leaned over the table, and whispered, "My parents aren't exactly normal either."

I know, I thought, but I didn't say it, didn't react at all.

"Really," Miyoko said, like maybe I didn't believe her. "Do you know any other family who fills all their outdoor pots and planters with fake flowers and plants every spring?"

I raised my eyebrows.

Miyoko continued, "Do you know any other family who keeps all their shoes lined up in the garage?"

I didn't know anybody else who did that, but I didn't know

anybody else who had as much white in their house—including some white carpet—as the Hoshis did either.

"At my mom's house, we keep our clothes in the laundry room," I offered.

Miyoko smiled. "There's only one thing we can do."

"What's that?"

"Eat cake."

It was really good cake.

I'd worked so hard on it and the cake had turned out so pretty—with pale purple icing and delicate, deep-purple violets. Beneath the light and airy icing, which I'd whipped until both arms ached, was a chocolate cake so moist, it clung to the fork. It melted in my mouth. Mmmmmm.

When the reception was almost over, I went upstairs with Miyoko to wait for her parents. I hadn't realized how noisy the reception was until I left it behind, and felt relieved by the quiet.

After Miyoko was gone, the thought of going back downstairs, facing the new Adams family, all the people who were here to celebrate it and expected me to celebrate, too—not to mention the noise they made—filled me with dread. So I didn't go back down to the party right away. Instead, I lay down on the wooden bench and watched dust float in the sunlight streaming through the stained-glass window above me. Because being lonely alone isn't nearly as bad as being lonely in a crowd.

Chapter 19

In the car on the way home from the wedding reception, Keene kept saying stuff to Mom like "Just think, Cecily, by tonight, we'll be on the beach, just the two of us!" and "I can't wait to have you all to myself for a whole week! I can't wait!"

Mom said stuff like "I know" and "Me too."

I felt invisible. Except to the sun, which blasted through the back window like it was aiming for me. There were no visors I could flip down in the backseat. Also, the air-conditioning never quite seems to make it back there. *Might as well get used to it*, I told myself, because somehow I knew I'd always end up in the backseat with Mr. and Mrs. Adams. So I sat there getting used to it, letting sweat—mixed with makeup—drip down my neck. We all went back to the town house to wait for Dad to come pick me up. I'd be staying with him and Suzanne over spring break while Mom and Keene went on their honeymoon and then moved Keene into our house.

Mom and Keene were still busy unloading their wedding presents when I walked into our cool, dark house. I looked around, wondering what the town house would look like nine days from now. Mom had already told me we'd have to move some of our furniture out to make room for Keene's stuff.

Suddenly, I loved the horse painting in our entrance hall, our comfy blue couch, our dining room table, and all the things we had that I'd never really paid attention to before.

I wondered what else would change, besides our house. Would our routine change? Would there be new rules—again? Would someone explain them to me, or would I have to figure them out by trial and error—again? How many errors could I make before Keene got really good and mad? What would happen then? Would Mom be on Keene's side? Always? Sometimes? All these thoughts and fears started spinning in my head, crashing into one another and exploding outward. I felt dizzy. And headachy. And sick.

But one thing was certain: Something had already changed. My house felt different, and I was suddenly uncomfortable in it.

Good thing my suitcase was already packed.

Despite the heat and humidity—it felt more like late May than April—I dragged my suitcase out onto our small front porch to wait for Dad, and then remembered I hadn't checked the mail today.

I was still standing by the mailboxes, shuffling through our mail, when Mom came up carrying a stack of boxes wrapped in whites and silvers and golds. "That reminds me," she said. "Fizzy, I need you to give your mail key to Keene."

"Why?" I said.

"Because we only have two keys . . . and Keene needs one."

"But I need one, too," I said, "to get my letter from *Southern Living*—it could come anytime in the next few weeks."

"We'll make sure you get your mail," Mom said in a very final way. Then she moved on: "Don't you want to change out of your dress?"

"No, thank you," I said. "I'm fine."

"Well, at least wash your face—your father won't appreciate the makeup and it's a bit of a mess now anyway."

I knew she was right about that, so I followed her into the house, angrily tossed the mail and my key onto the table in the entrance hall, and stomped off to the bathroom.

The cold water felt soothing on my face and neck. Even so, when I looked at my clean freckled face in the bathroom mirror, I almost *cried*. Freckles are just such a disappointment! I couldn't get out of there—away from the mirror—fast enough.

I bumped into Mom while fleeing the bathroom. "Sorry," I mumbled to my clicky wedding shoes.

Mom placed a hand under my chin and lifted my face. "Much better," she said.

When our eyes met, it took everything I had not to burst into tears. I was going to miss her, and it was more than just the honeymoon. I felt like I was going to be missing her for a very long time. I felt like I was losing her, like I'd already lost her—like my mail key.

Mom kissed me on the forehead, and then said, "I'm going to change."

I know, I wanted to say, but instead, I hugged her—tight.

Less than a minute later, I was back outside, on the porch, waiting for Dad to pick me up.

Keene came out to get the last load of presents.

"What if I need Mom?" I blurted.

"Hmmm?" Keene said, looking around like he'd forgotten I was there, as he mopped his sweaty forehead with a handkerchief.

"What if I get sick or something while you and Mom are away, and I need you to come home?" I asked. All he had to say was, *If you really need us, we'll be there.*

Keene shook his head and chuckled to himself as he stuffed the hanky back into his pocket—gross! "Oh no, you don't," he said, smiling like we were sharing some sort of inside joke. "We're not going to miss a minute of our honeymoon, not a minute, not for anything."

So something—big—had already changed. Even last week, my mom would've been there for me if I'd really needed her, but now . . . well, I was on my own, wasn't I? *Is this what happened to Zach?* I wondered. *Is this how he ended up "on his own"? And "slick"?*

"You'll be fine, Fizzy," Keene said as he trotted down the steps.

I decided that Keene Adams didn't know the first thing about being a good friend. *Good* friends don't completely ignore you, and they don't forget about you either. *Good* friends are there when you need them, no matter what.

Dad's car pulled to the curb.

I yanked my suitcase so hard that it tumbled down the steps and I all but ran for the car.

As we drove away, I watched Keene walk back up to the

house. He never even turned his head. I mean, I could've been kidnapped and he wouldn't have known or cared! *We're not going to miss a minute of our honeymoon, not a minute, not for anything,* I heard him say again in my mind, and I believed him. I really did.

"You clean up good," Dad said in a teasing voice as he glanced over at me.

Usually, he said this to me on Sunday mornings, when I came downstairs dressed for church. And usually, I answered him, "Yes, sir, I take a bath once a week, whether I need it or not." Then we'd both laugh. (Don't worry; I really take a bath every day. Really.)

But today I didn't respond, just stared out the window without really seeing anything.

"So your mom and Keene tied the knot, huh?" Dad said.

"Yes, sir."

Silence.

"Fizzy?"

"Yes, sir?"

"You okay?"

"Fine," I said, because I knew Dad couldn't help me with any of this, not even if he wanted to. And anyway, I knew he didn't want to. Dad didn't know Keene. And he didn't want anything to do with Mom, just like she didn't want anything to do with him. That's why they'd divorced. That's why they had *un*tied the knot, because they didn't want to be tied together anymore.

Of course, there was still a little knot holding them together: ME! I would always be the knot that tied two people who really, really wanted to be untied. When I realized this, I wondered how they could even love me. I wished I could untie them and let them go, because I loved them and I really did want them to be happy. I really did. But I didn't know how to do that.

I squeezed my eyes shut like I could make the truth go away by refusing to look at it.

"Are you tired?" Dad asked.

If I couldn't untie him, I figured the least I could do is keep my Mom-problems to myself. "Yes, sir," I said. "I'm fine, just tired."

"Me too," Dad said.

Chapter 20

I saw Suzanne's stomach before I saw her. She was lying on the couch looking like she might be trapped underneath it.

"*Wow*," I said. I didn't mean to. It just slipped out because, great gravy, that thing had gotten *huge*! It's true that Suzanne was going to have a baby . . . although, come to think of it, she had been eating a lot of ice cream lately. A lot. For a few seconds, I wondered if maybe Suzanne's belly was really just full of Choc-o-Chunk—her favorite ice cream—because if so, that belly was a real accomplishment and . . . an *expense* (food is money, Mom says). But I decided there was probably a baby in there—swimming in melted ice cream.

"I know," Suzanne said. "Wow is right. I've really popped! I'd get up but . . ." She looked at her belly, then at us, and with her eyes she seemed to ask, *How?*

"No, no, you just relax," Dad said. "I'll get dinner started."

I turned to stare at Dad like he was an alien, like he'd said, "I'm from the planet Crotuplkniat," instead of "I'll get dinner started."

"What?" Dad said when he noticed me looking at him like that.

"Nothing," I said quickly. I risked a quick glance at Suzanne.

She smiled and winked at me like we shared a secret, and I guess we did: We both knew that Dad didn't cook.

"Maybe Fizzy could help you," Suzanne suggested.

I nodded.

"Sure," Dad said. "What do you feel like having?"

Suzanne thought about it and then said, "Chili."

Dad's eyes widened. "Suzanne...Suze...honey, it's eighty-five degrees outside."

"Soooooo?" Suzanne said, drawing the word out, making it sound like a dare.

Dad cleared his throat. "So chili will be perfect...won't it, Fizzy?"

"Perfect," I repeated, wondering who in the world this guy was. I mean, he looked like my dad, and he sounded like my dad, but he sure didn't *act* like my dad.

Like I said, for starters, Dad didn't cook—had never even *tried* to cook as far as I knew. Also, I'd never known him to back down on a dare or an argument or anything like that. And lastly, if Dad thought it was ridiculous for his family to eat chili in April, I would've bet that nobody in his family would be eating chili. Period. But suddenly, I had no doubt that we would all be eating chili on this hot April evening, including Dad. Weird.

"Um, I just have to change my clothes," I said.

"You look pretty!" Suzanne called after me.

You should've seen me without the freckles, I thought.

• • •

I made chili for dinner that night, using steak cut up into little pieces instead of ground beef. Dad even added a cup of red wine when I asked him—the alcohol cooks out, but the layer of flavor remains. Unfortunately, he also added a little too much salt.

"That's okay," I told Dad. "Just peel a potato and drop it in the pot."

Dad got the potato and washed it, but then he hesitated. "Are you sure about this, Fizzy? I've never had chili with potatoes in it."

I turned from the stove. "You won't tonight either. The potato's just going to soak up some salt, and then we'll take it—and the extra salt—out of the pot."

Dad didn't ask any more questions after that. He did whatever I said, which was almost as much fun for me as the cooking. At one point, I was tempted to say something like, *Now hop three times on your left foot,* but I didn't because I wanted him to trust me in the kitchen.

I suddenly felt nervous when we were all seated at the table for dinner. I watched as Dad took his first bite of chili, and hoped he wouldn't say, *What did you put in this?* I knew from past experience—with Mom—that when he asked that, what he really meant was, *What did you put in this . . . to ruin it?* This seemingly innocent question had started many fights between my parents—fights that, in the end, weren't really about food at all.

But Dad smiled at me and said nothing.

Then Suzanne tasted my chili and said, "Oh my gosh! This chili is so good, I could *cry*!" And then tears filled her eyes.

I didn't know what to say: *Thank you? You're welcome? I'm sorry?* I looked at Dad.

He gave me a thumbs-up, so I figured everything was okay.

And it was. The chili was such a success that Dad and I made dinner together every night while I was at his house. And by that I mean *I* made dinner every night and—when he got home—Dad supervised my use of all major appliances. (Parents don't forget things like coming home to fire trucks with red flashing lights in the driveway.) I didn't mind. I love making dinner, any dinner, under just about any circumstances. I *missed* making dinner.

I hadn't gotten to do much cooking at Mom's lately because Mom had been doing most of the cooking. Keene would mention that he remembered eating—and loving—a certain something when he was a kid, and the next thing you knew, Mom was making that certain something, and had invited Keene to dinner. To make matters worse, I didn't usually like the certain something. *At all.*

But even when Keene hadn't been at our house and I'd been allowed to cook, my options were very limited due to the Wedding Diet that Mom was on—the Wedding Diet consists of baked, broiled, or grilled chicken or fish with steamed vegetables or salad.

But Suzanne wasn't on the Wedding Diet; she was on the Pregnant Diet. On the Pregnant Diet, you can eat anything you want—apparently. And what Suzanne wanted was rich, heavy foods and sauces, with complex flavors that take time to develop. So making dinner for Suzanne and Dad took hours, but it was my favorite part of being at their house.

The rest of the time was sort of boring. Dad went to work while Suzanne stayed home, decorated the baby's room, ate, and napped.

I watched a lot of TV. Well, except for on my last day, which was a Monday—an in-service day at school, for teachers only. On that Monday, Suzanne sort of went crazy over the baby's room. First, she decided the room wasn't clean enough—it looked clean to me. Then, she decided the color of the walls was all wrong and that we were going to repaint. Today. Right now, as in, "Fizzy Russo, turn off that TV and get in here!" It was sort of scary.

Even scarier, my dad left work in the middle of the day to buy paint and bring it to us. Now, my dad never leaves work in the middle of the day. But there he was, standing in the kitchen, in the middle of a Monday.

"You bought 'Garland Green,' right?" Suzanne said. "'*Garland* Green'?"

Dad nodded, but he looked anxious as he pried open the paint can to show Suzanne the muted kale-green color.

She said, "Oh yes, that's much better. I don't know why we didn't go with that in the first place."

Dad exhaled. "I've got to get back to the office."

I followed Dad out to his car, where I informed him, "That paint is the same color that's already on the walls in the baby's room."

"No, it's a different green," Dad said.

"But it looks *exactly* the same!" I insisted.

Dad ran a rough hand through his hair in a frustrated gesture. "I know," he finally admitted. "But please don't say that to Suzanne, okay?"

I shrugged.

"Okay?" Dad demanded.

"Okay, okay—good gravy!"

Dad nodded once and got into his car, while I stood barefooted on the warm concrete, shaking my head. *Who are you?* I thought as I looked at him. *And why are you wearing my dad's face?*

I found Suzanne upstairs, already rolling the new paint onto the walls in the baby's room. There was an extra rolling brush for me, so I picked it up and started painting. Every few minutes, Suzanne gave me instructions on painting, and I tried hard to do what she said. I mean, you never know what a crazy person might do if you make a mistake, right?

After a while, I asked, "How come you didn't ask Aunt Liz to decorate the nursery?"

"I wanted to do it myself" was the answer.

We did some more painting.

Then I asked, "How come we don't know whether you're having a boy or a girl? My math teacher's pregnant and everybody knows she's having a girl."

"I don't want to know," Suzanne said, still painting.

"Why not?"

Suzanne stopped and turned to look at me. "There are a

few very precious moments in life, Fizzy, moments you never get to have again, you know?"

"Um, not really."

Most grown-ups probably would've said something like, *You'll understand in time.* But one of the things I liked about Suzanne was that she hardly ever said stuff like that to me. Instead, she tried to explain things so I could understand them *now*—I'd have to remember to add that to her Like List.

"The day you get engaged, for instance," Suzanne said, "your wedding day, the day you give birth to your first child—those are moments you never get to repeat in life, not really."

"Okay," I said, "but I still don't understand why we can't know whether the baby's a boy or a girl."

"Because I want that moment to hold as much joy and surprise as possible—like Christmas morning. Do you understand?"

"I think so . . . yeah . . . you don't want it to be like Christmas when I was seven."

"What happened when you were seven?"

"I was out shopping with my . . . well, anyway, I was out shopping."

"With your mom?" Suzanne guessed—and she seemed okay with it.

"Yes, ma'am, and I fell in love with this beautiful ballerina doll, but it was too late then to be adding stuff to my list. I knew it was too late, but all of a sudden, I didn't want anything else—just this ballerina."

Suzanne nodded.

"So my... um, mom said she'd think about getting the doll, but I kept worrying about it and wondering if she'd gotten it yet—I was afraid the ballerina would be gone when she went back to the store."

"So what did you do?"

"I started sneaking around the house looking for my doll, and sure enough, I found her hidden away. But that wasn't enough."

"Why not?"

"Because then I wanted to play with her. I mean, she was *right there* in my house with me, and nobody was even playing with her!"

Suzanne laughed.

"So I snuck and played with her every chance I got, and by the time Christmas came, I was tired of the ballerina—that Christmas wasn't as exciting as the other ones."

"Precisely!" Suzanne said, picking up her rolling brush again.

I smiled and picked up my brush, too. "Well, I hope I get a half sister instead of a half brother."

"What?" Suzanne said.

Stupidly, I repeated, "I hope I get a half sister instead of a half brother." When I looked over at Suzanne, I knew I'd made a mistake.

"Fizzy," she said, frowning, "this baby is not a *half* anything. Calling the baby a *half* makes it sound like a lesser person, a lesser family member. This baby is a whole person, a whole family member, *your* sister or brother."

I got so nervous, I dropped my paintbrush and a little paint splattered onto the new rug.

Suzanne said a curse word.

She never cursed, so I was sure I'd heard her wrong.

But then she said it again, *three* times!

As Suzanne crouched to clean up my mess, I backed slowly out of the room.

I felt sad and worried that Suzanne and Dad were both losing their minds.

Chapter 21

Daffodils, tulips, pansies, forsythia, and cherry blossoms—there were explosions of yellow and pink and purple everywhere as Dad drove me home to Mom's. But even so, lots of ladies were already out planting more flowers in preparation for the Kentucky Derby—which is when people from all over the world descend on our city wearing their finest clothes and fanciest hats, to watch a two-minute horse race. The flower fireworks made me feel hopeful and happy for some reason.

But then I guess I must've sprung a leak because the closer we got to the town house, the less happy I felt. For a minute, I imagined our car leaking happiness all along the road, like oil—except that I'm pretty sure happiness is yellow, like butter. In cooking and eating, happiness *is* butter. And sugar.

"Fizzy?" Dad said. "Is something wrong?"

"No, sir," I said.

When we came to a red light, Dad turned and stared at me.

I looked out the window some more.

As the light turned green and we began moving again, he said, "Tell me what's going on."

I shrugged one shoulder.

"Tell me," he commanded.

"I don't know," I said. "I guess it's just that everything... every*one*... is changing."

"I'm not," Dad said quickly.

I turned to stare at him. "Are you kidding? You've changed more than anybody."

"What're you talking about? I haven't changed a bit."

I rolled my eyes. "You clean, you cook, you leave work in the middle of the day to buy paint that you already have, and you run to the store in the middle of the night to buy Choc-o-Chunk."

Dad smiled. After a minute or two he said, "Uh... pregnant women can be a little demanding and unpredictable sometimes."

I was really surprised—and relieved—to hear Dad say this. "I thought it was just Suzanne," I confided.

"No," Dad said. "The chemicals in a pregnant woman's body go a little haywire. Add to that the fact that they're hungry and nauseous, often at the same time. They're tired, but uncomfortable, so they don't sleep well. It's all those things and more—it's an exciting but also frustrating and frightening time in life."

"So you're scared of Suzanne?" I said.

"Of course not," Dad said. He chuckled to himself. "Well, maybe a little, but only in her current condition."

"Me too," I said. "I'm scared of her, too—she cries over chili... she goes crazy over paint..." *And she cusses*, I thought, but I didn't tell Dad what Suzanne had said, and I hoped she wouldn't tell him what I'd said either—about having a *half* sister or brother.

Dad laughed again. And then I did, too. I laughed even more when I remembered the plastic snake coiled under Suzanne's neatly folded pajamas, inside the overnight bag she'd packed for the hospital—for when the baby was coming.

For a few minutes, I felt better. Then I arrived home.

At first, I thought I'd walked into the wrong house, but then I saw Mom.

"Oh, Fizzy, I missed you so much!" she gushed, rushing to hug me.

I hugged back but didn't say anything. I was still busy gawking. I couldn't believe how different everything looked. The couch had been moved to one side and a big puke-colored recliner chair sat right smack in front of the TV, where *I* usually sat! I couldn't believe Mom had allowed this. Was it possible she hadn't noticed the ugly chair yet?

Mom pulled back from the hug and looked at me. "Well?"

"Where're all our photo albums?" I asked, noticing an empty-looking bookshelf that held a few knickknacks, but used to be jam-packed with family photo albums.

"Oh, I'm sure they're here somewhere," Mom said vaguely, working to maintain her smile.

"Where?" I said.

She lowered her eyes and her voice. "In a box, in storage—they're fine, Fizzy."

"I want them."

Mom turned and glanced at the stairs, and then said softly, "All right, I'll get them for you . . . later . . . on one condition."

I looked at her.

"They can't come out of your room," she said.

"Why not?"

"Shhh—because it would hurt Keene to see pictures of me as another man's wife—knowing and seeing are two different things."

Does it hurt him to see you as the mother of another man's child? I wondered. But I didn't ask. I figured I already knew the answer because I already knew that I wasn't exactly a cherished member of the Adams family. I'd known that I was Russo leftovers long before now. What I hadn't known was that in addition to the fact that the Russos and I were officially no longer part of Mom's family, we were also going to pretend that we'd *never* been a family—because that's why we were hiding the photo albums, right?

And it wasn't just the albums. I already knew from experience—with Dad and Suzanne—that I couldn't say things like "Hey, Mom, remember that time you and Dad and I went to Myrtle Beach?" over dinner with Keene. Because that would make Keene feel excluded, like an outsider—like me—which I can tell you *is* upsetting.

Ten years of history, gone, never to be mentioned again, I thought, but all I said was, "I understand."

Mom instantly brightened. "Good. Now, what do you think of the house?"

I thought I hated it, but knew I couldn't say so. "Um . . . you didn't redo *my* room, did you?"

Mom's face fell. "No—why? Don't you like what we've done?"

"Sure," I said, remembering: *It's best for everybody if you just say whatever the adults want to hear.* "Um . . . did I get any mail?"

Mom smiled knowingly. "Not yet, but I'm sure you'll hear from *Southern Living* soon—don't worry."

I spent the rest of the afternoon and most of the evening, too, in my room, having a little pity party for myself: no mail from *Southern Living*; a whole decade of my life and my family erased like the big fat mistake it was; and a house that was barely recognizable as my home. Well, at least my room still felt like mine.

When Mom called me down for dinner and I learned that we were just having smoked turkey sandwiches and chips, I asked if I could eat in my room.

Mom stopped assembling sandwiches and turned to look at me.

I looked away.

Keene pushed off from the counter he was leaning against and said, "C'mon, Cecily. Why not? It's been a long week for everybody."

"All right, all right," Mom said, and then she went back to the sandwiches.

I never took my plate back to the kitchen that night. I never went back downstairs for any reason. If Mom and Keene noticed, they didn't seem to mind. They were probably busy kissing each other, I told myself. Yuck.

When I knew I couldn't put off unpacking any longer, I swung my suitcase up onto my bed and unzipped it. A small box wrapped in yellow polka-dotted paper with a smushed white bow sat on top of my clothes in the suitcase. I sat down on the bed and opened it. Inside the cardboard box, I found a black velvet jewelry box and one of dad's business cards: ROBERT S. RUSSO, D.M.D. I turned the card over. On the back, Dad had written:

> *Remember that you have a heavenly Father, too, and He is always with you.*
>
> Love, Dad

The velvet box held the tiniest, most delicate gold cross necklace I'd ever seen.

I was struck with all kinds of thoughts and feelings. At first, they floated down out of nowhere like multicolored sprinkles: *Plink. Plink. Plink.* But then the sifty lid popped off, and they poured down fast and heavy on me. Too heavy.

Surprise and gratitude and even a little happiness struck first. Then anger hit me, along with the thought *This is not enough.* Even so, I was relieved that someone seemed to understand how worried and alone I felt. But the fact that that someone was Dad—who ordinarily wouldn't shop, buy me a gift for no reason, or write me a note—told me three things: 1) He was worried about Keene's presence; 2) he was worried about his own absence; and 3) he was worried about me.

Dad's worries made mine seem justified, which made me even more worried—downright scared.

Still, I put the necklace on and resolved never to take it off.

Then I put on a favorite old T-shirt to sleep in. When I remembered Keene was staying, I decided to sleep in some old shorts, too. And just as I'd thought, none of it was enough. I didn't feel any better—or any less scared.

Chapter 22

I felt sick and nervous about coming out of my bedroom the next morning. I felt *really* nervous about using the bathroom, which we now shared with Keene. So, I locked the bathroom door and hurried, not just to get out of the bathroom, but to get out of the house.

When I passed Keene in the upstairs hallway, he sort of grunted at me. I figured he agreed: I should get out of the house as soon as possible. As for me, well, *I* thought Keene should put on a shirt.

I felt much better outside. The sky was blue; the sun was shining; there was a cool breeze; and the birds were singing to each other, as if to say, *Look what a pretty day! Nothing bad could happen on a day like this! I know! I know!*

The longer I was outside and the farther I walked, the better I felt. Maybe Coach Bryant was right—maybe fresh air and exercise really did help. When I started walking up the hill on Dahlia Drive, I thought of Zach. He was probably waiting for me on his porch, so I picked up my pace. But when I reached the top of the hill, I could see that Zach's porch was empty. I slowed way down, taking tiny baby steps, to give him time to come out.

I'd never really looked closely at Zach's house before. It was made of rock and somehow seemed older than the houses around it. At first glance, it seemed smaller, too. But when I looked at the sides of the house, I could see how far back it went. It wasn't any smaller than the other houses; it just didn't want to be a show-off. I liked that about Zach's house. I decided his house was my favorite.

I walked the rest of the way to school as slowly as I could, all the while hoping Zach would eventually catch up to me. He never did, but that wasn't the worst of it.

I was late. Again. The hallways were all but empty when I entered, except for Mrs. Sloan, the—gypsy—guidance counselor. She saw me right away and made a beeline for me, her gold-coin belt jingling with every step.

"Hi, Fizzy," she said, smiling as if I were the person she'd most wanted to see this morning.

"Hi."

"Looks like you need a tardy slip."

"Yes, ma'am," I said to my shoes.

"I could help you with that."

Relief mixed with gratitude flooded my heart as I looked up at her.

"Tell you what," Mrs. Sloan said. "You come to my office and chat with me for fifteen minutes or so and I'll give you a tardy slip—an *excused* tardy slip."

I thought about this and then said, "But we'd just be chatting, right? I mean, it won't be like... *counseling*."

"Right. We're just two friends catching up. Come on," she said, and I followed Mrs. Sloan to her office.

She pulled out a chair for me at the little worktable.

I just stood there, looking at it. "Um... isn't that where you do the counseling?"

Mrs. Sloan smiled. "Sit anywhere you want, Fizzy."

Unfortunately, I didn't think I'd be allowed to sit behind Mrs. Sloan's desk, so I had to sit at the little table. I had to. Mrs. Sloan sat down there, too.

"So how was your spring break?" Mrs. Sloan asked.

"Fine."

"Did you go anywhere?"

"To my dad's house."

"How was that?"

"Fine."

"And now you're back at your mom's house?"

"Yes, ma'am."

"How's that going?"

"Fine."

Mrs. Sloan nodded, and then she seemed to think for a while. I waited. *Tick-tock.*

Mrs. Sloan laced her hands together and placed them on the table. "Fizzy, I'm curious," she finally said. "As a counselor, I can't help but wonder what you have against counseling."

"Nothing," I was quick to say. "I think counseling's great. It's just that *I* don't need it."

"Because you're fine," Mrs. Sloan said.

"Right."

"Do you think talking with me would somehow make you *less* fine?"

I sighed. "What do you want me to talk about?"

"What do you like to do when you're not at school?"

"I like to cook," I offered.

"Fantastic," Mrs. Sloan said. "What do you like to cook?"

"Anything. Everything."

"That's wonderful. Do you cook for your family?"

"Sometimes."

"What do they like for you to cook?"

"Well, my stepmom likes my chili. A lot. My dad likes any kind of dessert, but especially my banana pudding. My mom likes my red wine vinegar chicken, and her boyfriend—I mean, her . . . um, husband . . . likes my lasagna."

"Wow. You *do* cook everything," Mrs. Sloan commented. Then she asked, "Did your mom remarry recently?"

"Yes, ma'am."

"How do you feel about that?"

Sick. Scared. Displaced. Lonely. My nose felt pinchy and my eyes watered, but still, I said, "Fine. I want my mom to be happy and she is, so that's good."

Mrs. Sloan reached behind her, plucked a tissue from the box on her desk, and held it out to me.

I blinked back the tears and sniffed. "No, thank you. I'm fine."

Mrs. Sloan withdrew the hand with the tissue in it. Her

eyes pleaded with me. "Please, Fizzy, just talk to me. I can see that something's bothering you. What is it? You can tell me."

"No, thank you," I repeated.

"I've already told you that you don't have to worry about being polite here, so . . ." She shook her head and gave me a look, as if to say, *What's the problem?*

I sniffed again. "My mom says that when you talk about your own family behind their backs, it says more about you than it does about them—it tells people you're disloyal."

Mrs. Sloan leaned back in her chair and thought about this. "I think that depends on where you are, who you're talking to, and why. Talking to a counselor or a close friend, in private, isn't the same as discussing personal matters at the local beauty salon for anyone to hear. We aren't here to gossip or put others down or even prove them wrong."

"What are we here to do?" I asked.

"We're just here to talk through anything that might be on your mind, because two minds are better than one, right? Two minds are twice as likely to come up with a solution—to anything—right?"

"Right," I said, and then I looked at the clock. *Six more minutes.* "Two minds are better than one. So, what's on your mind?"

Mrs. Sloan laughed a loud, throaty laugh—it actually startled me. Then she said, "Well, let's see . . . I'm a little worried about my cat. He's sick, so I had to drop him off at the vet this morning. He hates the vet. And he hates *me* when I take him to the vet—he turns on me—see where he bit me?"

I looked at the little teeth marks on the flap of skin between Mrs. Sloan's thumb and forefinger. "Yikes. What's your cat's name?"

"Judas." Mrs. Sloan got up, went to her desk, scribbled out a note, and handed it to me—even though I'd only spent ten minutes with her.

I stood. "Thank you."

"I want you to think about what I said, Fizzy. I know you're fine. And very loyal. And talking with me doesn't change any of that. Okay?"

"Okay," I said.

"Come see me anytime you feel like talking. Anytime at all."

When I walked into my math class that afternoon, my desk had moved. All the desks had moved, having been rearranged into columns instead of rows. Each desk had a slip of paper taped to the top right corner, with a list of students' names next to their class period. After I found mine—which was nowhere near Miyoko's—I scurried to read the labels on the desks in front of me and behind me.

The desk in front of me belonged to Brian "Breakfast" Orr, while the desk behind me belonged to Mara "Motor-Mouth" Tierney. I thought of Brian as Breakfast because all he ever wanted to talk about was breakfast—what he ate that morning, what he was thinking of eating tomorrow morning, and occasionally, he might ask about *your* breakfast. If you weren't enthusiastic enough about breakfast, then Brian's face would turn beet red and he would say, "Breakfast is the most important meal of the day, you know!" But I was more enthusiastic about food than most, so we got along okay.

As for Motor-Mouth Mara, well, I think her name pretty much tells you what you need to know. What it doesn't tell you is that if you try to ignore Mara, she'll take her index finger and tap your shoulder urgently until you're tempted to turn your head and bite it off. Personally, I'd only been a victim of this twice, but twice was enough.

Since Mara hadn't arrived yet, I sat down at her desk, leaned forward, and stretched out my arm. Darn. My desk was within tapping distance of hers.

"What are you doing, Mara?" someone said, from right behind me. It didn't really sound like a question, more like an accusation.

I froze, but allowed my eyes to wander to the two big feet stuffed into black high-heeled shoes that came to a stop beside me.

"I *said* what are you doing?"

"Nothing," I said quietly, my eyes still on those strange feet—the skin was very white and the bulging veins beneath it were very blue, like miniature blueberries lined up beneath wax paper.

"Well, stop it," the owner of the blueberry feet said, moving past me, headed for the front of the room.

Quickly, I got up and moved to my desk.

As soon as I sat down, Brian—who'd arrived while I was trying out Mara's desk—turned around in his seat and said,

"Man, you wouldn't believe the eggs Florentine I had for breakfast this morning!"

I gave him a weak smile—I was still thinking about Mara and looking for Zach, whom I hadn't seen all day.

Ms. Mini-Blueberry Feet stood and said, "Good afternoon, class."

A hush fell over the room.

She continued, "You'll be pleased to know that your teacher, Mrs. Carter, has given birth to a healthy baby girl. She'll be on maternity leave for the rest of the year, so I'll be filling in for her."

I gave Miyoko a worried look across the room, which she returned.

Our new teacher went on: "Please note where you're sitting. This is your assigned seat and if you're not in it when the second bell rings, you *will* be counted absent. Even if you're here."

Everybody looked around, uncertain, as if to say, *Can she really do that? That's not fair!*

"Now then, my name is Mrs. Ludwig and this will be my first time teaching school. I'm a retired police officer from Washington State."

"Cool!" someone said.

"Have you ever been shot?" someone else said.

"As a matter of fact, I have," Mrs. Ludwig said, coming out from behind her desk. She bent and pointed to a hole in her calf, just below the hem of her skirt.

Maybe *hole* isn't the right word. It was more like a big chunk missing out of her leg. But whatever you want to call it, it wasn't pretty. I wanted to look away, but for some reason, I couldn't.

I was glad when Mrs. Ludwig moved back behind her desk and sat down, because I couldn't concentrate on a word she was saying with that hole staring out at me.

It turned out that Mrs. Ludwig had told us to stand up and introduce ourselves.

When Brian sat down, I stood up and said, "My name is Fizzy Russo—"

"It most certainly is not," Mrs. Ludwig snapped.

Huh? I just stood there, confused.

"Your name is Mara Tierney," Mrs. Ludwig said, "and you should know that I don't appreciate pranks, Mara."

Mara's hand shot up in the air.

Mrs. Ludwig ignored her and said to me, "Please start over, Mara."

"Um . . . I'm not Mara," I said.

But Mrs. Ludwig was already shaking her head.

"I'm Fizzy," I said. "Really."

"She is," Mara said. "*I'm* Mara."

"Out in the hall, both of you," Mrs. Ludwig barked, rising from her desk. "*Now.*"

Somehow, Mara straightened things out in the hall. I mean, I guess she did. She must have because I'm pretty sure I didn't say anything. I was too busy staring at the hole in Mrs. Ludwig's leg.

Until I realized that Mrs. Ludwig was staring at me. "Ma'am?" I said.

"I *said*, why were you sitting at Mara's desk?"

I glanced over at Motor-Mouth and realized that there was no nice way for me to answer that question—at least, not truthfully—so I said, "I don't know."

"I know why," Mrs. Ludwig said.

I was so relieved that Mrs. Ludwig understood that I started to smile, but Mrs. Ludwig wasn't smiling. Mrs. Ludwig looked like she wanted to arrest me. That's when I knew she didn't *really* understand. And she didn't like me.

"Um . . . have you talked to Mrs. Warsaw, the principal?" I asked, thinking maybe Mrs. Warsaw had told Mrs. Ludwig not to like me.

"What? Of course I've talked to Mrs. Warsaw. She hired me."

Hired you not to like me? Yep, it was just as I'd thought.

When class was over, Mrs. Ludwig stood by the door watching us file out into the hallway one at a time. When I passed by her, she handed me a sealed envelope and said, "Please give this to your parents, Fizzy."

Chapter 23

"Well?" Miyoko said, squinting against the afternoon sun as we stood together on the sidewalk, just out of view of the school. "What does it say?"

I folded Mrs. Ludwig's note back up and put it in my pocket. "It says that I created a disturbance during math class today, that learning time is very valuable, and suggests a family discussion regarding these important matters."

Miyoko made a worried face and sucked in air through her teeth.

I understood. I felt worried sick at the thought of "a family discussion regarding these important matters," because a family discussion might include Keene.

The breeze picked up and the sun moved behind a cloud as we started walking. "Are you going to your aunt Liz's today?" Miyoko asked. "I'd love to go with you again sometime."

It *had* been a bad day and I was tempted, but, "No," I decided out loud. "I've got too much homework, especially math."

"I know! I can't believe it! What kind of substitute assigns this much homework?"

"Officer Ludwig," I said.

Miyoko giggled.

I wasn't nearly as happy.

Miyoko tried to cheer me up. "Have you heard from *Southern Living*?"

"Nope."

"Maybe today's the day."

"All I can think about is the letter I already have—from Officer Ludwig."

Miyoko nodded her understanding.

I sighed and said, "Sing 'Nenneko yo' again." "Nenneko yo" was a Japanese lullaby that Miyoko had told me her grandmother used to sing to her.

Miyoko smiled knowingly and began to sing:

> *Nenneko, nenneko,*
> *Nenneko yo,*
> *Oraga akabo no*
> *Neta rusu ni,*
> *Azuki wo yonagete,*
> *Kome toide,*
> *Aka no mamma e*
> *Toto soete,*
> *Aka no ii-ko ni*
> *Kureru-zo!*

It was a really good lullaby, I decided as we walked, because it made me feel a little calmer.

"What does the rest of it, besides *sleep, baby child*, mean?" I asked.

"The singer is saying that she will make some red beans, rice, and fish, and feed it to the best babies," Miyoko said. "Now you sing it with me."

I was just getting the hang of it when Miyoko suddenly stopped singing.

I turned and followed her line of vision to her house, which had just come into view: Miyoko's mom was running around their front yard, yelling like a maniac.

Miyoko and I slowed and exchanged nervous looks.

It was only when we got a little closer that we realized Miyoko's mom was chasing a dog. A dog that didn't belong to the Hoshis. A dog with a shoe in its mouth.

"Oh . . . it's okay—this happens all the time," Miyoko said. "My dad forgets to close the garage door and our neighbor's dog comes over and chews up our shoes."

I couldn't help smiling as I watched Mrs. Hoshi chase the big shaggy dog. When I realized Miyoko was looking at me, I sucked in my cheeks to stop the smiling. "Sorry. But it *is* sort of funny."

"I know—but it makes my mom *really* mad. I better go help her."

I nodded.

Miyoko took off running, but before she could get to her mom or the dog, disaster struck: Mrs. Hoshi's wraparound skirt had come untied, so that when she slipped and slid down the small hill in the Hoshi's front yard, the skirt got left behind. When Mrs. Hoshi stood up—without her skirt—she was wearing big, white granny panties.

Once on her feet, Mrs. Hoshi raised her fists, threw her head back, and let out a long, hair-raising scream. A face appeared in the front window of a house across the street. (*Note to self: If you're ever standing in your front yard wearing only big, blinding-white bloomers, do not scream.*)

Mrs. Hoshi turned and ran back up the hill, snatched her skirt out of Miyoko's hands, and then stomped into the house.

The sight of Mrs. Hoshi's backside caused my mouth to fall open: There was extra padding sewn into the butt of her panties. *Why?* I wondered. *Did she do really rough . . . sitting?*

Miyoko turned and offered me a sickly smile, shrugged, and walked toward the dog, who now stood in a far corner of the yard, still holding the shoe in his mouth and wagging his tail like, *Get-the-shoe is best the game ever! Let's play some more!*

I figured maybe I should help Miyoko—mostly because I felt bad about having seen her mom's weird underpants.

The dog ran around the side of the house and disappeared. Miyoko followed him. I followed Miyoko.

As soon as I rounded the corner, a hand clamped down on my arm and held it. It was Miyoko, with her back flat against the house, so that no one—like Mrs. Hoshi—could see her out of their windows. Miyoko's other hand covered her mouth as tears streamed from her eyes and her shoulders bounced up and down. I felt even worse about what I'd seen then—obviously, Miyoko was humiliated to the point of tears.

I looked away, trying to give her privacy, but Miyoko removed the hand from her mouth and said, "I'm not laughing *at* her . . . I'm . . . I'm . . ." Then she fell to pieces again—she was laughing!—and had to slap her hand back over her mouth.

Relieved, I smiled and tried to help her out: "So . . . you're laughing *with* her? But do you really think she's laughing?"

"She should be—that's the funniest thing I've ever seen! Did you see her butt pads? My mom has a really flat butt, so I guess . . ."

After that, we both had to clap our hands over our mouths. By the time we got control of ourselves, both the dog and the shoe were long gone.

I walked over to my backpack, lying in the grass, and then remembered: "Hey, Miyoko."

She turned.

"My mom said you could spend the night with me on Friday."

"I have to ask my mom," Miyoko said. "*Later.*"

I nodded my understanding, hefted the backpack onto my shoulder, waved, and started for home.

I remembered what Miyoko had said at my mom's wedding: *My parents aren't normal either.* They weren't. They really weren't. *Does* anybody *have normal parents?* I was beginning to wonder.

I laughed all the way home—every time I'd start to settle down, I'd picture Mrs. Hoshi in her Big Booty Judy Bloomers, screaming at the sky, and start up again.

Chapter 24

When I got to my house, I went to the mailbox before I remembered I didn't have a key. That's when the laughter officially died. *I'm probably lucky to have a key to the house!* I thought as I unlocked the front door. I stepped inside and was again struck by the feeling that I'd barged into the wrong house—the one with the pukey recliner.

I kicked off my shoes, dropped my stuff by the front door, and went to the kitchen to call Mom.

I got her voice mail: "You've reached Cecily Adams..."

I was thrown off balance in a way that was a lot like the wrong-house feeling. I had to think for a second. I thought I'd called Cecily Russo, and that's who I really wanted... but she was gone now and I was stuck with Cecily Adams. My throat tightened with this thought, but at the beep, I rasped, "Hi, Mom. I'm home," and hung up.

I was sitting at my desk, trying to do my math homework (translation: staring into my math book with a blank sheet of paper in front of me) when I heard Keene come home from work. I didn't go downstairs to say hello and he didn't come upstairs to say anything to me either. *So what?* I asked myself. Whatever.

A few minutes after Mom got home, she came into my room holding up the shoes I'd kicked off at the front door.

I just blinked at her.

"Fizzy, you've *got* to stop leaving your shoes by the door—bring your things up to your room, like I asked you to."

"Okay," I said, "but . . . why?"

"Because Keene doesn't like it," Mom said. "His pet peeve is floors—he always wants the floors to be clear and clean."

"Oh, good gravy," I mumbled.

Mom shot me a warning look.

I went back to my math book.

"How was your day?" Mom asked.

I gulped and touched the pocket that held Mrs. Ludwig's note. "Um . . . fine."

"It doesn't sound like it was fine to me," Mom said. "What's wrong?"

"Nothing," I insisted.

Mom waited a few seconds and then said, "*Elizabeth. Ann. Russo.*" This pretty much meant, *Spill it.*

I took a deep breath and blurted, "My new math teacher sent home a note saying that I created a disturbance in class today."

"Did you?"

"I didn't mean to."

"Would you like to explain?"

I explained.

"All right," Mom said. "Sounds like it was just a misunderstanding."

"So . . . um . . . we don't need to have 'a family discussion,' right?"

"Hopefully not . . . unless it happens again."

My shoulders sagged with relief. "I could make dinner," I offered.

"No, that's okay," Mom said. "I have a certain something planned already."

"A certain something" is what Mom calls a meal she's planning as a surprise for Keene. In case I wasn't perfectly clear before: I *hate* all the meals that Keene loves.

For a few minutes, Mom moved around my room, putting my shoes in my closet, smoothing the clothes that bubbled up out of drawers, closing the drawers, and straightening. All the while, I continued staring into my math book—and not writing anything.

Before I knew what was happening—and could stop her—Mom was out in the hallway, hollering down the stairs, "Keene, Fizzy needs you to help her with her math homework, okay?"

"Okay," he hollered back. "Tell her to bring it down."

I looked at Mom, and with my eyes, I tried to say, *Please tell me you did not just do that.*

She smiled proudly at me like, *Problem solved.*

My stomach flipped.

Even so, I brought all my math stuff downstairs to the dining room table—what choice did I have?—where Keene was already seated. The first thing I learned is that it's hard to learn when your brain is busy worrying about what your teacher thinks of you and what he might think of you if you don't learn fast enough, or at all.

"Fizzy, are you listening?" Keene asked.

"Yes, sir."

He shook his head and then repeated some math mumbo jumbo.

I thought about how all math might as well be a foreign language... other than French, because I would understand at least some of that.

Keene stared at me.

I stared blankly at the fractions in my math book without moving.

Keene sighed.

"Never mind," I tried. "I'll just—"

"No," Keene interrupted. "Wait. Let me think of another way to explain it."

I waited—in a squirmy, uncomfortable way.

"Fizzy, how many slices can you get out of a pie?"

I snapped to attention. "What kind of pie?"

"It doesn't matter," Keene said.

I gawped at him. *Of course* it mattered! "So, then, if I promise to make apple pie for dessert next week, you won't mind if I make mincemeat pie instead?"

"Apple. It's an apple pie," Keene said, catching on quickly. "How many slices?"

"I'd say six good-size slices."

Keene nodded and quickly drew a pie with six slices. He colored in one slice, and said, "One slice is..."

"One-sixth!" I said, the concept of fractions making some sense when applied to food.

"Good, Fizz," Keene said. "I'll be right back."

Keene was gone long enough that I began to wonder if he'd forgotten about me.

But he came back, carrying a tray with a plate of sliced apple, a glass of water, a measuring cup, and a box of graham crackers. Keene picked up all the apple slices and put the apple back together. "This is one whole apple," he said.

I nodded.

He opened his hands a little so that I could see he'd cut it into eight slices. "If you eat a piece, that's—"

"One-eighth!" I said.

Keene used the water and the measuring cup to show me that I'd been working with fractions successfully for a long time—in my cooking. Then he used the graham crackers to make a number line, showing me another way to look at fractions.

I finished my math homework in twenty minutes flat and Keene checked it for me—perfect.

"Thanks," I said, when he handed back my paper.

Keene nodded in a tired way.

"And . . . I'm sorry. About my shoes."

"I know you are," Keene said quietly, getting up from the table.

As I set the table for dinner, I couldn't help wishing that Keene had responded, "It's okay," instead of "I know you are." Because what did that even mean? *Yes, I know, you are sorry— you are a sorry human being if I ever saw one?*

• • •

Keene was washing his hands in the kitchen while Mom bent to pull a pan of sweet potatoes out of the oven. She made a whimpering sound.

Both Keene and I turned at the sound.

Mom set the pan down and stuck a finger in her mouth.

"Did you burn it?" Keene asked.

Mom nodded.

"Let me see," he said. Keene carefully rinsed Mom's burn in cool water, wrapped it ever so gently in gauze, and dosed her with aspirin. When he was done playing doctor, he bent his head and kissed her wounded hand in a way that struck me as . . . *tender*. He loved her, I realized then, not in words, but in actions.

When we were all seated at the dining room table, I said, "What is that smell?"

"Brussels sprouts," Mom said cheerfully, passing me a big bowl filled with stinky-ness and what looked like tiny heads of lettuce.

"No, thank you," I said, pushing the bowl away and coughing a little.

Mom raised her eyebrows at me. "Fizzy, I've worked hard on this dinner. The least you can do is taste it—you can't know that you don't like something unless you taste it."

I couldn't believe it. I couldn't believe I was going to have to put something that smelled so awful on my plate, right next to my *food*! But Mom and Keene were waiting, so I spooned one of the little green things onto my plate and coughed some more.

Neither Mom nor Keene seemed to notice my cough. But *I* noticed, as we passed food around, that Mom only took a few of the stinky little green things, too.

The sweet potatoes were the good news. The bad news was that there wasn't anything else on the table that I liked even a little. So I ate my sweet potato and then just mushed and pushed everything else around on my plate.

Mom and Keene talked, mostly about work.

When there was a pause in conversation, I said, "May I please be excused?"

Mom inspected my plate and said, "Not until you try the brussels sprouts."

I looked right back at Mom's plate and said, "You haven't tried yours."

Mom looked like she wanted to say something but instead she speared the green thing with her fork and put it in her mouth.

I watched her chew, and I have to say, it didn't look like Mom enjoyed brussels sprouts.

Even so, she swallowed, touched the corners of her mouth with her napkin, forced a smile, and said, "Your turn, Fizzy."

Okay, now I'm not going to describe the gagging and coughing I did in detail, because . . . gross. No, let's just say that in my opinion, brussels sprouts not only smell like feet; they taste like feet, too!

"You may be excused, Fizzy," was all Mom said.

· · ·

When I was back in my room, I added to Keene's Dislike List:

10) *Likes brussels sprouts.*
11) *Took my mail key!*
12) *Has a weird thing about clean floors—eye roll.*
13) *Brought the ugliest chair in chair-history into our living room!*

Then I added to Keene's Like List:

5) *Is good at math.*
6) *Loves my mom.*

Chapter 25

I smiled a little secret smile when I saw Zach waiting for me in the distance, on Wednesday morning, as usual.

The first thing I said to him as he headed down his front walk to meet me was, "Do you like brussels sprouts?"

"I don't know," Zach said. Then he grinned his crooked grin. "Why? Were you thinking of cooking some for me?"

I felt myself blush. "*No.*"

Zach smiled some more.

I took off walking and he scrambled after me.

"Admit it," Zach said, coming up alongside me. "You missed me."

I glanced at him out of the corner of my eye. "When?"

"I knew it," he said, nodding and smiling to himself. "You missed me. *Bad.*"

I laughed.

"I missed you, too, Fizzy," Zach said. "But I had to be in court yesterday."

"Because you really are a violent maniac?" I teased.

Zach laughed. "Nah, because my grandmother wanted to legally adopt me."

"Oh. Where's your mom?"

The smile faded as Zach shrugged. "Who knows? Haven't seen her in a few years and I don't remember my dad at all."

"I'm sorry, Zach," I said.

"S'okay," Zach said easily.

I didn't know what else to say, so I just gave him a sympathetic look.

"Don't look at me like that," Zach said.

"Like what?"

"Like you feel sorry for me—I hate that—I'm *fine*."

"Oh yeah, I know that," I said quickly, because I understood: He just wanted to be a normal kid, like everybody else. I wanted that, too—for myself and for Zach.

"And Gran's fine, too," he said. "She's tough, but fair."

"So . . . then it's good that your grandma wants to adopt you?"

"Very good," Zach said. "It means I'll never have to go back to living with some other family or in a group home for foster kids—because I have my own family, my own home."

My heart ached for Zach. It must've been awful not having anyone who wanted him, even temporarily. And I knew how hard it was to live with one stranger, let alone a house full of strangers.

"Fizzy?"

"Yeah, sorry—so how'd it go in court?"

Zach smiled, remembering. "Gran stood up in front of the judge and the lawyers and everybody and said that she's capable of caring for me, that she wants me, and . . ."

Silence. I risked a quick glance at Zach out of the corner of my eye; he lowered his head.

We kept walking.

Finally, he lifted his chin and continued in a hoarse voice, "She said that I'm a good boy and she loves me. In front of everybody."

I didn't know why tears gathered behind my eyes, but I blinked them back, sniffed, and said, "So you'll live with your grandma from now on?"

"Yeah," Zach said. "I've been here, with her, almost a year now."

"That's great," I said. "I've been in the valley for almost a year now, too—hey, do you feel sick when your alarm clock goes off in the mornings?"

"No. Why? Do you?"

"*No*," I said defensively, and then I admitted, "Sometimes... only sometimes."

"Maybe you're not a morning person."

"Maybe," I agreed.

"Or maybe you need a new alarm clock," Zach said. "But me? I wake up feeling all right. Living here with Gran's been the best—it's better than foster care for sure."

A big dog barked ferociously from a fenced backyard as we approached. "He kinda reminds me of our new math teacher," I warned.

Zach grinned. "Can't wait."

In math class that afternoon, when Buffy entered the room, she smiled and trilled her fingers at Zach.

Zach smiled back.

Then, as soon as Brian sat down at his desk, he turned around in his seat and said, "Man! I gotta tell you about the omelet I had for breakfast this morning—diced tomatoes, feta cheese, fresh parsley! Man! I think it was the best omelet I've ever eaten."

I leaned over my desk, and said, "Let me ask you something."

"Yeah?"

"Do you like brussels sprouts?"

Brian looked at me like he was having serious doubts about my taste. He shook his head and said, "Aw, man . . . *brussels sprouts? For breakfast? Disgusting.*"

I was about to say that brussels sprouts are disgusting for breakfast, lunch, *and* dinner, but before I could, Brian—who was clearly deeply upset by the revolting turn in our conversation—turned back around.

I was still thinking about brussels sprouts when Mrs. Ludwig started our math lesson. Which is how I missed the exact page number we were supposed to turn to in our books. I looked around and tried to catch someone's eye so I could ask.

Zach's eyes were already on me, from the desk beside mine, across the aisle.

"What page?" I whispered.

"Two thirty-one," Zach said.

"Thanks."

"Who's talking?" Mrs. Ludwig boomed, looking around the room.

I sat up in my chair and stared straight ahead without blinking.

Finally, Mrs. Ludwig gave up, turned around, and started writing on the board.

I peered over at Zach.

He grinned, mouthed the words *Who's talking?* and made a funny face.

A giggle escaped me.

Mrs. Ludwig spun around. "Fizzy Russo, was that you?"

My face and ears went hot.

Just as I opened my mouth to apologize, Zach's math book slammed to the floor.

Mrs. Ludwig turned her evil eye on Zach. "Now that you have our attention, Zach Mabry, is there something you'd like to say?"

Zach shrugged his shoulders and gave Mrs. Ludwig a bored look.

What's wrong with you? I thought at him. *Say what she wants you to say!*

"Get your things, Zachary, and move to the table at the back of the room, please," Mrs. Ludwig said.

Zach looked around, then pointed to himself, and said, "You talking to me?"

Mrs. Ludwig stared him down.

"Because my name's not Zachary," Zach informed her. "It's Zachariah."

"Get your things, *Zachariah*," Mrs. Ludwig growled.

Zach did as he was told and Mrs. Ludwig moved on with our lesson on word problems.

I hate word problems. *Hate* them. Word problems are the brussels sprouts of math. Yuck.

When Mrs. Ludwig finally started winding down, I snuck a peek back at Zach: He'd built a fort out of the old textbooks stacked on the back table, and all I could see was the tip-tip-toppy of his blond head moving around in there! I don't know why it struck me as so funny, but it did. I laughed out loud.

Mrs. Ludwig whirled around. "Fizzy Russo!"

I tried to stop laughing. But the harder I tried, the harder I laughed. I laughed so hard my whole body shook. I couldn't stop. Not knowing what else to do, I squeezed my eyes shut and covered my face with both hands. And then, somehow, I laughed myself right out of my chair and onto the floor.

The whole class roared with laughter.

Mrs. Ludwig barked, "Out in the hall!"

I glanced back at Zach as I got up off the floor.

Zach rose from his seat just enough so that I could see his face above the fort. He winked and smiled at me. "Don't worry," he whispered. "I'll be there in a minute."

"Zachariah Mabry!" Mrs. Ludwig thundered. "You move those books this instant!"

Zach must've given a little push, because the fort suddenly toppled over and books crashed onto the table and floor.

Again, the class erupted in laughter.

Zach was right. Mrs. Ludwig glared at him, pointed at the door, and shouted, "You! Out in the hall, too!"

Zach followed me out into the hallway, where I stuffed my hands into my pockets and studied my shoes.

Mrs. Ludwig followed Zach. She came to a stop just two feet from us, towering over us, hands on her hips—looking very scary. "Zachariah, you *will* pick up every single one of those books."

Zach shrugged.

"*And* you will come to my classroom to stay after school every day for the rest of the week."

Zach smirked. "Mrs. Wilcox has already requested the pleasure of my company after school today. But I'd be happy to join you tomorrow and for the rest of the week—Monday, too, if you like. Thank you for inviting me, Harriet."

My jaw dropped.

"Do not dare address me by my first name again," Mrs. Ludwig warned.

I looked at Zach as if to say, *Have you lost your mind?*

"And you!" Mrs. Ludwig said, shaking a finger at me. "Get a hold of yourself, girl!"

"Yes, ma'am," I said.

"Now back to class, both of you."

When math was over, I paused in the doorway, waiting for Mrs. Ludwig to hand me the sealed envelope I was sure she had for my parents.

But she just gave me a cold look and said, "You're holding up the line, Fizzy."

Out in the hall, Mike Anderson gave Zach a high five and said to me, "Best math class ever."

I smiled. For a few minutes, I felt like I'd gotten lucky. Then it hit me: Mrs. Ludwig probably didn't write to my parents because she planned to *call* them. She might as well have punched me in the stomach—what if Keene answered the phone?

Chapter 26

After school, I found Aunt Liz in her warm, buttery kitchen. Her eyes smiled when she saw me. "I was just making some more Benedictine for you and Miyoko," she said.

"Thanks, but Miyoko couldn't come today. She said she'd see you next time."

"More Benedictine for you," Aunt Liz said as she flipped the switch on her food processor and a loud humming filled the room.

I smiled because I'd arrived at just the right time—Benedictine is best straight out of the food processor.

When Aunt Liz turned the processor off, the phone was ringing.

"Hello," she sang into it happily, just before her face changed. "Okay," she said seriously. "Okay. . . . Uh-huh. . . . How long? . . . Okay, I'm on my way."

I frowned. "I'm on my way" didn't sound like good news for me—or my Benedictine.

"That was your dad," Aunt Liz said, tearing off her apron. "Suzanne's having the baby!"

"Now? But it's not time yet," I protested.

"The baby doesn't know that—it's coming," Aunt Liz said. "C'mon, Fizzy. We've got to go—I'll drop you at home on my way."

"Why?" I whined. "Why can't I go with you?"

Aunt Liz grabbed her purse. "Because we don't know how long it'll take. The baby could be born in a few hours or it could be tomorrow. C'mon! We've got to go!"

"You promise you'll call, right?"

"Yes, I promise!"

As usual, my eyes went straight to the pukey recliner when I walked into the town house. The thing that was *un*usual was that Keene was sitting in it.

"Um . . . hi," I said, feeling awkward, like I should've knocked or something.

"Hi," Keene said.

I waited for him to say more.

He didn't. He just stared at me.

Had Mrs. Ludwig called? I wanted to ask, but didn't want to have to explain, so instead I said, "Any mail?"

Keene shook his head.

"Any calls?"

He shook his head again and continued to stare at me.

I decided Keene was probably busy making a mental list of all the things he hated about me—my meatball head on top of my toothpick body, the bump on my nose, my freckles, my raggedy old backpack. I couldn't blame him. I hated all those things, too.

"Well, um . . . I guess I better call Mom and let her know I'm home."

Keene nodded.

I dropped my backpack and kicked off my shoes before I remembered Keene's pet peeve. Then I picked everything up and carried it into the kitchen with me.

"Mom, Keene is *here*," I whispered urgently into the phone.

"Yes, he lives there now, Fizzy," Mom said matter-of-factly.

"Oh. Right." Then why did it still feel like there was a guest in the house?

After I'd told Mom that school was fine and I was fine—in fact, about to become a big sister any minute—I put the phone back. Then I thought about the guest-y feelings some more. I remembered Keene's outburst during The Meat Loaf Dinner: "I am not a guest!" he'd insisted angrily.

It was then that I realized he was right: Keene wasn't the guest.

I was the guest! In my own house!

I looked at Genghis every five or ten minutes and waited for the phone to ring all night, hoping it would be Aunt Liz—and *not* Mrs. Ludwig. But it never did. So after I'd—likely incorrectly—finished my math homework, I went downstairs and sort of hovered at the edge of the living room.

Mom and Keene were watching *Survivor Steve*, who was talking about primitive man's survival instincts: "Primitive man did not use a pillow. He listened for danger with both ears while he slept . . ."

I waited for a commercial and then gave a little cough.

Keene looked at me like, *Darn. Are you still here?* and muted the TV.

"Mom, I'm sure the baby's been born by now. Please drive me to the hospital, *please*," I begged. "You don't have to get out of the car or anything—you can just drop me off."

"No, Fizzy. I'm sorry, but you can't go wandering around a hospital by yourself. You'll just have to wait," Mom said. Then she went back to watching TV with Keene, just like she used to watch TV with *me*.

I took a bath and went to bed without saying good night to anyone, because somehow I knew that another interruption would irritate Keene—even more.

The next thing I knew, Mom was sitting beside me on the bed, saying lightly, "Fizzy . . . Fizzy, honey."

I opened my eyes and propped myself up on an elbow.

Mom switched on the lamp on my nightstand. "Your dad called."

I pushed the hair out of my eyes.

"You have a new baby brother. He came four weeks early but is going to be just fine. They named him Robert, after your father."

I nodded. "So you'll take me to the hospital now? Dad—or Aunt Liz—I'm sure somebody can meet me."

Mom smiled. "No, visiting hours are over—it's almost midnight. But I have a little something for you now."

That's when I noticed the mixing bowl in Mom's lap. I leaned over and peered in: some kind of chocolate batter and two spoons.

I had to hand it to Mom: She really surprised me sometimes. I scooched over in the bed to make room for her. Mom handed me the bowl and climbed in, under the quilt.

We ate brownie batter and talked about the new baby and what it means to be a big sister.

"As he gets older, your baby brother will look up to you," Mom said. "He'll want to do everything you do—so you'll need to set a good example."

I nodded.

"And be patient with him. Be patient when he keeps hanging around and you wish he'd go away. Remember that he does this because he worships you and wants to be just like you."

"I guess I could stand being worshipped," I said.

We laughed. We laughed a lot. I felt happy and comfortable.

I felt comfortable enough to risk asking, "Do you really like the pukey recliner downstairs?"

Mom laughed again. "No, but Keene loves it, and I love Keene. Love means compromising. Compromising means sacrificing. For love."

"Oh," I said. "I thought compromising meant meeting somewhere in the middle."

"It does."

"Well, that chair doesn't look like you met Keene in the middle," I said. "It looks like . . . a big, sick sacrifice."

Mom laughed some more. "We met in the middle, Fizzy. There are things of mine that Keene isn't fond of either."

Like me? I wondered, but I didn't say it.

Mom stayed in bed with me until the oven timer beeped downstairs. We decided to save the actual brownies for tomorrow, so Mom got up, tucked me in, and kissed me good night.

As I drifted back to sleep, I realized that for the first time in what seemed like a long time, I didn't feel like an interruption or an inconvenience, or a leftover or even a guest. Being home is a good feeling. But I knew it wouldn't last.

Chapter 27

I woke up to the sound of electronic screeching, thunder, and a heavy downpour of rain. I turned the sickening, screamy Genghis off, rolled over, and went back to sleep—because who in their right mind would get up and *out* in this kind of weather? It seemed like good thinking at the time, when my brain was still fuzzy with sleep—like my teeth.

Thirty minutes later, daylight must've registered with some primitive part of me that is still concerned with survival. Survivor-me apparently didn't like our chances of surviving another visit to the principal's office, so she alerted sleeping-me that there was a problem. I opened my eyes. My room was filled with dim blue-gray light, but I could see. I could *see*! "No, no, no! That *can't* be the time!" I told Genghis. Even if I started my morning with a run through the rain, I'd still end up in Mrs. Warsaw's office—for being tardy.

"It's all right," Mom said when I came barreling out of my room, into the hallway. "Keene's going to drive you."

I stopped short and blinked at her.

"He is always on time," Mom promised. "Oh, and I hid the shoes you left in the kitchen yesterday in the cabinet under the sink, if you need them." She gave me a look like, *Didn't I tell you not to leave your shoes out?*

I lowered my eyes and nodded.

I found Keene waiting for me in the kitchen, where I retrieved my shoes as quickly and as casually as possible. I slipped them on my feet and said, "Um... I'm ready."

"Don't you need to eat something?" Keene asked.

"Oh no—I don't eat breakfast."

Keene looked like he didn't believe me. "Cecily?" he called.

Mom practically came running. Of course.

"Fizzy says she doesn't eat breakfast," Keene said in a very snitchy way.

Mom glanced at me and then explained to Keene, "She's never cared for breakfast—it's fine. She eats a good lunch, don't you, Fizzy?"

I nodded, even though this wasn't entirely true. Granted, breakfast has never been my favorite meal, but I used to like oatmeal with maple syrup back when I lived at home—and Mom and Dad lived there, too. But now I wake up with the homesickness in my belly and eating—especially oatmeal—just makes me feel sicker. As for eating "a good lunch," I eat what the school cafeteria serves, so I guess it depends on your definition of *good*.

Keene gave Mom an unsure look, then shook his head, grabbed his keys, and said, "Okay. If you say so."

Keene didn't say a word in the car, so we just sat there in total silence—like two strangers seated next to each other on an airplane. Naturally, Keene's car was spotless—no straw wrappers or crumbs from fast food restaurants, no dirt on the floor mats, no dust in the vents.

As we rode, I thought about Zach. If he was still on his porch, I planned to lower the window and wave at him as we passed by, but Keene took a different route to school. One part of me hoped Zach had gone on to school without me, and the other part hoped he'd sit on his front porch waiting for me all day.

I thanked Keene for the ride when he dropped me off, and checked the clock as I entered the school hallway. Mom was right: Actually, I arrived at school seven minutes earlier than usual. Which just gave me extra time to dread math class.

At lunchtime, Miyoko and I couldn't find Zach, even though Miyoko said she'd seen him at school this morning. We decided he must be doing some kind of detention. I hoped to see him in math class—if I absolutely *had* to go to that.

I still didn't want to go to math that afternoon, because I figured the mere sight of me would remind Mrs. Ludwig to call my parents, but I went anyway—what choice did I have?

"I waited for you this morning," Zach said as soon as I sat down at my desk.

I smiled and said, "I knew it: You missed me. *Bad*."

The second bell rang and when Mrs. Ludwig entered the room, naturally, something horrible happened: a pop quiz on word problems.

To make matters worse, I was the last one out after class and I saw Zach and Buffy standing together in the hallway, talking. I did a double take and when I looked back at them, Buffy was performing a perfect hair toss. Zach rewarded her with his crooked grin. I felt more than just a teensy bit nauseated.

When the phone rang that night, I ran to answer it. But it wasn't Aunt Liz or Dad—or, thankfully, Mrs. Ludwig.

Instead, Miyoko's voice said, "Okay, I can spend the night tomorrow . . . on a few conditions."

"*Conditions?*" I repeated.

"Yeah," she said, and then she started running down The Conditions.

"Hold on," I interrupted. "I think I need a pen and paper."

I stood at the edge of the living room, waiting for a commercial, as Survivor Steve said, "All survivors have three things in common: the ability to accept their situation, adapt to it, and move forward—quickly. More on that when we return."

The commercial started: "Does your house smell like a litter box?"

Keene muted the TV, and he and Mom both looked at me expectantly.

"Um, Miyoko can spend the night . . . on a few conditions," I announced.

"*Conditions?*" Mom repeated.

I nodded and read my list aloud:

1) *Miyoko is not allowed to watch TV, play on the computer, or use any electronics, because she's on restriction for the B she got on her English paper.*

2) *Miyoko may go outside but we must stay together at all times, and we aren't to wander off.*
3) *If Miyoko injures herself in any way, she must call home immediately.*
4) *Miyoko is to be in bed no later than ten o'clock.*
5) *She must be up, dressed, packed, and ready to go on Saturday morning by nine.*

I looked up.

"Is that all?" Keene asked.

I nodded.

Mom turned to Keene. "What do you mean, 'is that all'? Isn't that enough?"

Keene shrugged.

Mom turned back to face me. "Honestly, Fizzy, I don't know whether to laugh or cry—for myself or for Miyoko." She shook her head. "I don't even know what to *think*."

"Please, Mom," I said, because I really wanted Miyoko to come.

"All right, I guess . . . but—did something happen when you were at Miyoko's house—was there some sort of *incident*?"

"No, ma'am," I said quickly.

"Did you use your manners?"

"Yes, ma'am, I ate tofu and broccoli at dinner and everything."

Mom did some thinking and came up with, "You didn't discuss private family matters, did you?"

"No, ma'am."

"Then why do Miyoko's parents seem to think you're being raised by wolves—or nobody at all? I mean, 'no wandering off'? And 'if Miyoko is injured'? What's that supposed to mean?"

"Um . . . Miyoko has a tiger mom," I informed her. I was tempted to add, *Also, she wears Big Booty Judy Bloomers* just so Mom would understand how NOT normal Mrs. Hoshi really is, but I didn't feel comfortable discussing panties in front of Keene.

Keene elbowed Mom to indicate that *Survivor Steve* was back on.

"Fine," Mom said. "But you better start thinking about what you're going to cook for Miyoko, because a girl who's been eating tofu deserves a decent meal."

"Yeah," I said. "With sugar. And butter. And salt. Wait—so I get to cook? For Miyoko?"

Mom nodded.

"Take your shoes up with you when you go," Keene said.

"Yes, sir," I mumbled.

I was halfway up the stairs when Keene called out, "Oh yeah, you got some mail today, Fizzy. I put it in your room, on your dresser."

My heart started hammering in my chest at the thought of mail—possibly from *Southern Living*!—and I rushed to my room. But it was only an advertisement, "a special offer" to subscribe to *Young American* magazine. I wadded it up and threw it in the trash angrily. Then I asked myself, *Why are you so mad?* Of course, I decided it was Keene's fault.

I didn't like it that Keene had come into my room when I wasn't here. And I didn't like it—still—that I'd had to give him my mailbox key. And my TV-watching spot. And my mom. And I *really* didn't like it that I was hardly allowed to do any cooking, and wasn't allowed to leave my shoes on the floor *as if I lived here!*

I searched my room for signs of snoopage. That's when I noticed a pair of panties lying on the floor, next to my hamper. Keene had seen my panties! On the *floor!* How awful! *At least they're normal panties and not Big Booty Judy Bloomers,* I told myself, but it didn't help.

I took a bath to try to calm myself. I wrapped my wet hair in a towel, put on my sleep T-shirt and shorts, and picked up my room so that nothing was left on the floor—except the furniture—and so that no one would see anything I didn't want them to see. Well, unless they were snooping, which would be wrong, very wrong—do you hear that, Keene Adams?

After that, I planned Friday's dinner: fried chicken with mashed potatoes and gravy, macaroni and cheese, buttermilk biscuits, and homemade peanut butter ice cream for dessert.

Then I got out my Keene lists. On my Dislike List, I added:

14) *Sneaks into my room when I'm not around—Suzanne's mom was right: There is nothing worse than a sneak!*
15) *Took my TV-watching spot!*
16) *And also MY MOM!*
17) *Makes me feel like a guest—in my own home!!!*

On my Like List, I—begrudgingly—added:

7) *Is punctual.*

But my Dislike List was still way longer than my Like list. *So there,* I thought.

Chapter 28

Miyoko loved my house. Well, okay, maybe not the house itself, but the absence of tiger mom inside it. And she loved the food. When we were all seated at the dinner table, the night of our sleepover, Miyoko closed her eyes, inhaled through her nose, and swooned a little. Keene said, "I just want you all to know that I'm willing to loosen my pants if I have to—for seconds." As she ate, Miyoko gasped and moaned with pleasure. Keene responded, "I know, I know," with a mouthful of food.

Miyoko said that being at my house felt like a vacation to her. I felt proud and told her she could come over anytime she wanted, and I didn't even think about ruining Miyoko's vacation spot by mentioning my guestness—or hers. I'd wanted her to feel at home and when she did, it gave me hope—hope that maybe I'd feel at home someday, too.

And hey, maybe someday I'd even get to meet my new baby brother.

But another three days passed, making it five whole days since my brother had been born, and I *still* hadn't seen him! I didn't know the first thing about him!

I was mad at Dad because he'd never called back to talk to me—or arranged for me to visit the baby and Suzanne in the hospital. I was mad at Mom because she wouldn't let me have my own phone I was sure Dad would've called if I'd had my own phone and he hadn't had to worry about getting stuck on the phone with Mom again.

And I was *really* mad at Aunt Liz, who hadn't been home when Miyoko and I had stopped by her house, and who hadn't called even though she'd promised.

When Mom gave me the message that Aunt Liz had—finally!—called, I decided to ignore her right back. Maybe I'd return her call in a few days. Or maybe not. Because it was obvious that when her new nephew had been born, Aunt Liz had forgotten all about her old niece, which was wrong, very wrong. And . . . well, I expected more of Aunt Liz, that's all.

I mean, ever since my parents' divorce, I felt like I'd lost a really important grocery bag, the one with all the important ingredients—for my life. Substitutions had then been made: new house, new neighborhood, new school, new friends, new stepmother, new stepfather, and now a new brother! These are all highly noticeable changes in the recipe of my life, which means they aren't good substitutions, because *good* substitutions aren't noticeable. But these were so noticeable, I felt like I'd been given someone else's ingredients, for someone else's life. But I just had to keep on living it—what else could I do?

And while I was living it, the one person I'd been able to count on—until now—was Aunt Liz. Through it all, Aunt Liz had stayed the same as always. Whenever she saw me, she gave me the big smile that reached her eyes—even when she was busy—instead of that tired what-now? look I usually got from everybody else. Aunt Liz was *always* happy to see me. And I was always happy to see her because when I was with her, I felt the same as always, too: comfortable and right at home—*wanted*—*loved*. Sometimes when I was with her, I even forgot how I'd lost so many important ingredients. But now . . . well, Aunt Liz was probably saving the Big Smile for Baby Robert. I wasn't sure how I'd stand it if she ever gave me the tired look.

Zach walked up the hill to meet me that Monday morning.

"You're going the wrong direction—school's that way," I said, pointing.

He grinned. "Wasn't looking for school."

I felt shy all of a sudden, so I just nodded and we started walking—toward school.

After a few minutes I asked, "Do you ever feel uncomfortable in your house?"

"Uncomfortable how?"

"I don't know, like you're intruding or . . . interrupting or . . . like maybe you're not supposed to be there." I was sorry I'd said it as soon as the words were out of my mouth. I risked a quick glance over at Zach while I waited for him to say something like, *No. That's weird. You're weird.*

Zach's face had gone all serious. "Not anymore," he said, "but I know what you're talking about: You feel like you don't belong."

"Yeah," I said, and I felt so relieved that someone understood, even a little bit, that I wanted to cry. But I didn't—and that's what's important.

"You belong," Zach said, staring off into the distance, his jaw clenching and unclenching. "With me, you belong. Okay?"

"Okay," I whispered, still trying not to cry. It took me the whole block to pull myself together. But I did it.

Just before math class ended that afternoon, Mrs. Ludwig handed back our word-problem pop quizzes. When she laid my quiz facedown on my desk, I knew I was in trouble. And I was right: When I turned my paper over, there was a big red D on it, next to the words *Parent Signature Required*. My stomach felt queasy as I considered my options. If I showed my quiz to Mom, I was sure she'd force Keene to help me with word problems. If I showed it to Dad, he'd be mad—plus, he was obviously too busy to help me with anything—or even remember that I exist.

My mood remained gloomy and doomy as Miyoko and I walked home together, even though it was a near-perfect afternoon, the sun moving in and out of poufy white clouds that made me think of meringue pies.

"What's wrong?" Miyoko finally asked.

"I got a D on my math quiz *and* I have to have a parent sign it."

Miyoko winced and sucked in air through her teeth.

"Thanks a lot," I said. "What'd you get?"

Miyoko ducked her head and admitted, "An A."

I slapped my forehead with the heel of my hand. "Right. Duh."

"But I have to get As. I *have* to, Fizzy."

I knew she meant because of her tiger mom.

"Hey," Miyoko said. "Maybe we could stop by Aunt Liz's house and she could sign your quiz for you."

I shook my head.

"She wouldn't do that?"

I shrugged. "Probably not, but I'd have to go to her house and ask to find out for sure . . . and I'm never ever going there again as long as I live."

Miyoko frowned. "Fizzy," she started, and I could tell that she was about to try to talk some sense into me. I hate it when Miyoko tries to talk sense into me, because she's usually right. I don't want her to be right; I mostly just want her to be on my side.

Someone yelled, "Hey! Fizzy! Miyoko! Wait up!"

We both turned.

Zach was jogging to catch up. When he slowed to a stop in front of us, Miyoko said, "You don't have to stay after school today?"

"Nah," Zach said, sounding a little disappointed.

Miyoko gave me a puzzled look; she'd heard the disappointment in Zach's voice, too.

"You don't *like* staying after school, do you, Zach?" Miyoko said.

"Why not?" Zach said. "Sitting in the air-conditioning doing homework that I'd have to do anyway beats sweating it out in the heat *and then* doing homework."

"So that explains it," I told Zach. "You said it's best for everyone if you just say whatever the adults want to hear, but you never say what Mrs. Ludwig wants to hear—because you *want* her to keep you after."

"Yeah, I guess it depends on what you're trying to accomplish," Zach said, smiling. "My social worker says that I use good and bad behavior equally to get what I want—she says that's my way of taking back some control over my life."

"Oh," was all I said.

Zach continued, "She also says that I have trouble with some authority figures... but I like old Ludwig... and she likes me."

"Ludwig does *not* like you," Miyoko informed him as we all started walking.

"She doesn't like me either," I added so he wouldn't feel as bad.

Zach grinned some more. "Y'all are wrong. She likes me."

"How do you figure?" I said, because I knew Ludwig didn't like *me*—and I didn't cause her half the trouble Zach did.

"Think about it," Zach said. "If she didn't keep me after school every day, she could go home. But she doesn't. She stays. With me. So she must really enjoy my sparkling personality."

Miyoko and I laughed.

Zach added, "*And* she drinks soda after school, which she shares with me—come to think of it, y'all are right: She doesn't like me; she *loves* me."

"I can't even picture that!" Miyoko said, still laughing.

Then I got serious. "Well, I'm not sure if Mrs. Ludwig likes you... but we all know Buffy Lawson definitely does."

Zach rolled his eyes. "So?"

"So you don't like her at all?" Miyoko said like she didn't believe him.

I *wanted* to believe him. But it was hard.

"Nah," Zach said. "I've known lots of Buffy Lawsons—there's at least one at every school. You two, on the other hand, are originals—much more interesting."

Miyoko and I both smiled at the compliment.

Zach stepped off the sidewalk, into his front yard.

Miyoko and I just stood there gaping. The bushes! Great gravy! The bushes around Zach's house had been cut down to sad little nubs. Their chopped-off limbs and leaves were scattered all over the place. I felt like I was looking at a botanical crime scene—a very violent one.

Zach followed my line of vision, looking behind him, and then said, "Oh. I did that. Yesterday."

"Why?" I asked. I mean, it just seemed like such an angry thing to do to bushes. After all, what could the bushes possibly have done to Zach?

"I wanted to play basketball with some neighbors," Zach said, "and my grandmother said I could, after I finished my chores. But when she told me what all I had to do, I knew I wasn't going to get to play—I knew it'd be dark before I was done. It's *always* dark before I'm done."

I just kept staring at those poor bushes. *How had I missed them this morning?* I wondered. Then I remembered: Zach had met me at the top of the hill and by the time we'd passed his house, I'd been busy with Operation Don't Cry.

"Gran should've just said no when I asked to play basketball, you know?"

"But how... *w-why?*" Miyoko stuttered.

Zach shrugged. "I had to mow the grass and trim the bushes... so I made sure the bushes wouldn't need trimming again for a long, long time."

"Did you get to play basketball then?" I asked.

"No, and now I'm grounded. Not that it makes any difference—I get chores when I'm good and chores when I'm bad," Zach said.

"You're like Cinderfella," Miyoko said.

Zach grinned. "Nah, Gran and I just have different life philosophies, that's all."

"What do you mean?" I asked.

"Gran grew up on a farm, so her idea of good parenting is no shenanigans and lots of chores," Zach explained. "I like lots of shenanigans and no chores."

Miyoko and I laughed.

The front door opened then and Zach's gran appeared. "Zach! Chores!"

Zach nodded and gave us a look like, *See? I told you.* "Later," he said.

Miyoko knelt on the sidewalk and retied a loose shoelace.

I waved at Zach's grandmother.

She didn't wave back, and when Zach passed her in the doorway, she smacked him on the back of his head.

Yikes.

Chapter 29

On Friday, I was *finally* going to meet my baby brother. Thankfully, nobody—like Keene—was home after school, so I was able to pack for the weekend at Dad's without feeling nervous and rushed.

A car horn honked outside, announcing Dad's arrival.

"Oh, good," Dad said when I opened the back door and hung my church dress on the little hook inside.

"Oh, good, what?" I said irritably, because I was still mad at him.

"Suzanne was hoping you'd bring that dress," Dad said.

I got into the car and pulled the seat belt over me.

"We're having a photographer come to the house this weekend to take pictures of the family."

I didn't respond.

"Ready?" Dad said.

I nodded.

As he steered the car away from the curb, Dad said quietly, "Your necklace looks nice on you."

"Thanks. Thank you. For the necklace . . . and stuff," I said. Then I turned to look out the window: The whole city was in bloom for Derby Day tomorrow.

"How've you been?" Dad tried.

"Fine," I mumbled.

"How's school?"

"Fine."

"Excited about meeting your new baby brother?"

"Yes, sir."

Dad nodded. "He's healthy, whole, and . . . *perfect*."

Perfect, I thought. *Unlike me. That explains a lot.*

It was almost six o'clock when Dad and I arrived home—because of all the extra Derby-weekend traffic—but Suzanne was still wearing her nightgown. She sat rocking the sleeping baby in front of the TV. I couldn't help noticing that she didn't look too happy.

"Colic," Dad whispered to me.

I nodded like I knew what that meant and stayed back.

"What's for dinner?" Dad asked.

For a split second, Suzanne looked like she wanted to kill Dad and eat *him* for dinner. But then she smiled sweetly and said, "You can make anything you want."

"Pizza it is," Dad said, removing his jacket and folding it over the arm of the couch.

Dad went to Suzanne, kissed her cheek, and then carefully lifted the baby out of her arms.

Suzanne stood and stretched. "I'm going to take a shower."

Dad nodded. "Come here, Fizzy," he said. "Sit down."

When I was settled in the rocking chair, Dad bent and placed the baby in my arms. "Make sure you support his head," he instructed.

"I will."

Dad stepped back and smiled at the two of us. Then he said, "I'm going to order the pizza. Stay right there while I'm on the phone—don't move."

I nodded.

Baby Robert was wrapped in a soft blue blanket. He was smaller than I expected, but heavier, too. He had lots of dark hair and a nose as cute as a cupcake! Also, he smelled really good—sweet, but not like cupcakes or cookies. I stared at him and breathed his milky sweet scent.

Baby Robert's blue eyes popped open. Somehow, I felt like I'd been caught doing something I wasn't supposed to be doing. For a minute, we just stared at each other. Then a deep, grapefruit-pink color appeared on his forehead, spread to his ears, and moved downward as he opened his mouth and started to cry. Really loud.

"Oh no . . . no, no, don't do that," I said. I tried rocking him, but he kept right on crying.

When Baby Robert's sweet smell was replaced by some other—hideous—smell, I nearly started crying myself. "Help! Help!" I shrieked.

Dad came rushing over.

"Take him! Take him!" I said, gagging.

"Shhhh, it's okay," Dad said, gathering the baby in his arms.

I didn't know if he was talking to me or Baby Robert; we were both pretty shaken up.

That night, I learned what *colic* means. It means the baby cries whenever it's awake. I wished we'd gotten a puppy instead.

• • •

It was still dark when Dad came into my room and woke me up. "The photographer will be here in an hour," he said.

I sat up and tried to make sense of what Dad was saying. Then I remembered the family portraits. "We're having our pictures made in the middle of the night?"

"It's morning," Dad said.

I looked at my still-dark window; it didn't *look* like morning.

"Sunrise and sunset provide the best natural light for pictures," Dad explained.

"Well, why didn't you pick sun*set*?" I asked. I mean, this was ridiculous.

"The baby does better in the mornings," Dad said.

"Well, *I* do better in the evenings," I muttered under my breath.

"Yes, but you're not a baby," Dad said, shooting me a warning look. "So get dressed."

Dad does better in the evenings, too.

The photographer, Raul, was a bossy man with a ponytail. He arrived at our house and set up his camera near the big sycamore tree in our backyard, which was surrounded by red coralbells that Mom and I had planted together when I was six. "Perfect!" he announced. "Come! Come!" He motioned to us with one hand.

Suzanne was put into position beneath the tree first. She held Baby Robert, who was still sleeping, in her arms.

When that was done, Raul said, "Now the father," and Dad

stepped forward. When the three of them looked absolutely perfect, Raul said, "And the daughter."

"Stepdaughter," I corrected, moving forward. I didn't know why I'd said it; I guessed I was still mad and it was still too early in the morning.

"Ah, I see," Raul said seriously.

For a while, Raul took pictures of all of us. When he finished, he looked up from his camera and said, "How about mother and child?"

After that, Raul took pictures of just Suzanne and Baby Robert. "Beautiful," Raul told Suzanne. "And now the three of you?"

"Please?" Suzanne said.

There was no question who Raul meant by "the three." I knew who he meant and Dad did, too. In his defense, I have to say that Dad looked a little torn, a little sad-ish maybe, but still, he took his place under the sycamore with his new family.

And that is how I ended up standing alone on the back porch, watching someone else's family have their picture made, against an orange-sherbet-tinged-with-raspberry-sorbet sky, wishing I had never said "stepdaughter"—and wanting to pinch Baby Robert, just a little bitty bit, because he'd kept us all up most of the night, and now that *we* had to be up, he was sleeping like an angel.

As I stood there breathing the scent of fresh mint that we'd planted next to the porch, I couldn't help remembering all the Derby parties we'd had here. Mom had always served iced tea with sprigs of mint in them on that day—to make them special.

Then I remembered that today was Derby Day. Not that it mattered. Anymore.

After that, Suzanne asked Raul to take pictures of me alone. *Why?* I thought. *I know I'm alone. I don't need pictures of it.*

Something must've registered on my face, because Suzanne hurried to explain, "I thought a nice photo of you might make a good Christmas gift for you to give to your mom."

Right. Whatever, I thought, but I stepped off the porch and headed for the tree.

When we were done taking pictures, I went back to my room to change clothes. Then I sat down and made a list of all the people I was mad at:

1) *Suzanne*
2) *Dad*
3) *Aunt Liz*
4) *Mom*
5) *Keene*
6) *Mrs. Ludwig*

Since it was already shaping up to be a bad day, and since I was already mad at him—so he might as well be mad at me, too—I decided to show my math quiz to Dad after lunch.

"A *D?*" Dad said, taking off his reading glasses to give me his glare of disapproval in full force.

"Yes, sir," I said, "and I need you to sign it."

Dad put his reading glasses back on and looked over my

paper some more. Finally, he shook his head and said, "I'm sorry, but I can't sign off on work like this, Fizzy."

"But you *have* to," I said, trying not to panic, trying not to think about how Keene Adams was going to have to teach me word problems when he could hardly stand to look at me.

Dad held my paper out to me. "I'll sign it when you correct it."

"But I don't know how," I pleaded.

"Your teacher didn't teach you how to do this?"

"Ummmmm . . . she tried, I guess."

"And?"

"Well . . . see, she has this hole in her leg—it's *really* distracting . . ."

It turned out that Dad couldn't understand how a bullet hole in your math teacher's leg could prevent you from learning word problems. But he *could* understand word problems. Actually, he was a word-problem whiz. The trick, Dad told me, was to underline the important words—the math words—and to ignore all the other unnecessary information.

By Saturday evening, I was pretty good at word problems, too. I felt like I'd accomplished an impossible feat, like I'd mastered a foreign language in a single day—and I sort of had, because before today, to me, word problems had read something like this: *At 4:00 p.m., Sally gets on a train traveling 35 miles per hour. She has three pencils and two pens. How many waffles can she make before polar bears become extinct? Answer: Pink. Because it doesn't rain on Mars.* But not anymore! When I brought my last practice paper downstairs for Dad to check, I found him sitting up on the couch, sound asleep. His head had lolled forward and his chin rested on his chest.

I turned to go—back to my room.

"Fizzy," Dad said.

"I didn't mean to wake you up," I said, *even though* you *woke me up in the middle of the night*, I thought.

"You didn't—I wasn't sleeping—just resting my eyes," Dad said, holding out his hand for my paper.

I gave it to him and waited while he looked over each problem, pen poised, ready to make Xs.

Without making a single mark, he gave the paper back. "Very good."

"Thanks," I said.

I was almost to the kitchen when Dad said, "By the way, Aunt Liz told me you wanted to come to the hospital when the baby was born and I appreciated that. But . . . Baby Robert was premature, so there were extra precautions and tests, and I had my hands full—with Suzanne and the baby in the hospital, Suzanne's parents flying in from Virginia at the last minute and staying here, and my practice and patients still needing me."

I nodded my understanding: *You were too busy with your perfect new family to be bothered with your old leftovers. Yep, got it.*

Chapter 30

On Sunday, Dad dropped me off at Mom's right after church so that I could attend Keene's family reunion with him and Mom. I didn't want to go, but since nobody had asked me, I figured I didn't have a choice.

Mom met me at the front door. "Hi, sweet pea," she said, hugging me.

"Hi."

"Did you have a nice weekend?"

"Yes, ma'am."

"I'm glad. Listen, we'll be leaving for the reunion in about an hour and you'll want to change clothes—it's just a backyard picnic."

I nodded.

"Oh, and this came in the mail for you yesterday," Mom said, grabbing a blue envelope off the table in the front hall.

I took the envelope and immediately recognized Aunt Liz's lovely, loopy handwriting on the front. I started for the stairs.

"Don't forget your suitcase," Mom said helpfully.

"Right. Wouldn't want to leave anything out on the floor," I muttered, "because it might remind somebody that I live here."

"Fizzy, I don't like your tone," Mom said.

I plodded up the stairs, thinking, *Just add it to my list.*

I closed my bedroom door behind me, plopped down on my bed, and tore open the envelope. There was a card inside. On the front was a picture of a sad dog looking out the window. Inside, the card read: *I miss you. Love, Aunt Liz*

Truthfully, I missed her, too. But then I imagined Aunt Liz giving me the tired look; my throat tightened and my nose stung. I decided not to think about Aunt Liz any more right now. I replaced the card in the envelope, dropped it into a drawer, and shut the drawer—quick.

Keene's sister, Hadley, lived across town in a historical house with a big backyard. The yard was swarming with picnic tables, lawn chairs, and people—people I didn't know. Oh, sure, I'd met a few of them at Mom's wedding, but I didn't really know them. I didn't really *want* to know them, which turned out to be a good thing, because they didn't seem like they wanted to know me either.

Well, except for Hadley herself, who pinched my cheeks, hugged me, and told me I should call her *Aunt* Hadley in a voice so sugary, I could practically feel a cavity coming on.

"*Aucune possibilité,*" I said, which meant "no chance" in French. Did I mention the woman actually *pinched my cheeks*? Also, she wore way too much perfume.

"Oh! She's bilingual!" Hadley gushed. "Fabulous!"

Mom shot me a warning look.

I ignored Mom and said to Hadley, "May I use your restroom?"

In the bathroom, I tried—unsuccessfully—to wash the lingering scent of Hadley's stinky perfume off me.

I spent the rest of the time hidden under the long branches of a willow tree, sitting in a woven blue-and-white lawn chair that made the backs of my legs itch, reading a cookbook I'd borrowed from the kitchen. That is, until I heard my name.

". . . very sweet of Keene," said a woman's voice, oozing with sympathy, "to accept a child that isn't his—he lets Fizzy live with them and everything."

I felt my whole head go hot, like someone had lit a fire under my chin. I was burning with anger and hurt and shame—not to mention stinky perfume—but mostly anger. I mean, for someone to actually say that Keene *lets* me live in my own home? I was there first! As far as I was concerned, it was very sweet of *me* to accept a man that wasn't my father and let him live with *me*!

I slammed my book shut, got up, and went looking for Mom.

The closer I got to the grill, the more people I saw eating cheeseburgers. Then I spotted Mom, who was just about to take her first bite of one.

I marched right up to her and blurted, "I need to go home."

Mom removed the cheeseburger from her mouth, set it down on her yellow plastic plate, and exchanged uh-oh looks with Keene.

Uh-oh is right, I thought. I was prepared for an argument. In fact, I was prepared to repeat what I'd just heard, and *not* quietly.

But I didn't get an argument. Instead, Mom handed her

plate to Keene, wiped her mouth, and said, "I just have to get my purse, okay?"

Mom thanked Hadley for having us and told her that I wasn't feeling well.

Hadley gave me a pouty look and said, "Bless your wittle face."

I gave her the squinty eyes.

Keene stayed at the reunion while Mom and I went home.

Once we were in the car, on our way, Mom said, "Do you want to talk about it?"

"No, ma'am," I said as a lone tear slipped out of my eye and streaked down my cheek. I swatted it away with the back of my hand.

Mom nodded and kept her eyes on the road.

By the time we arrived home, I understood that I truly was a guest in my own home. That the roof over my head wasn't really mine, and that the man who considered it his was unrelated to me, not responsible for me, had no obligation to me, and didn't love me—he'd said so himself. And people knew it, felt sorry for Keene, and thought he was "sweet" to tolerate me—*sweet!*

Knowing this made me want to leave, but where else could I go? Maybe I could stay with Dad and Suzanne. Maybe poor Suzanne would be "sweet" and tolerate me. Maybe. But what if Suzanne wasn't feeling so sweet? What if she was tired and cranky from all that awful colic? Or what if she let me move in, but then she *got* tired and cranky and changed her mind later?

What if Keene changed his mind? Could I quit school, get a job and an apartment? I would do it now if I could—that really seemed like the best option for everybody—but I was pretty sure there were laws against it. So what if neither Keene nor Suzanne could stand me for one more second?

Would I be . . . *homeless*? Sent to live with strangers in foster care, like Zach?

I didn't know, but I figured I'd better clean up my act. No more Ds. And no more Bs either. I had better start being *perfect*—like Miyoko.

I guessed it was because we were alone in the house—for once!—that Mom finally decided to let me have my own sort of family reunion.

I'd just crawled into bed when she came into my room, lugging a box marked *Photo albums*.

"Thank you," I said.

Mom set the box down on the floor and nodded at it. "Make a place for this in your closet or something—I don't want them left lying around."

"Yes, ma'am," I said.

But still, Mom stood there wringing her hands and looking worried.

I sat up. "I understand that you don't want Keene to see them. I'll make a place for them right now."

Mom exhaled. "Thank you," she said, and then she left, closing my door behind her.

I worked for a solid hour, making room for the box in my

closet. But before I slipped it into its new space, I pulled out an album and sat down on my bed with it.

On the very first page was a photo of Mom, Dad, Gamma and Grampa Russo, Aunt Liz, Uncle Preston, and me, taken at our house on Christmas morning. It was the last Christmas morning we would all be together, but I'd had no idea and it showed: The ten-year-old girl in the snowflake pajamas bubbled over with joy. I actually remembered *being* her, feeling exactly the way she felt: full to bursting with the happiness of all my favorite people and foods and things. And then I realized I would never feel that way again.

Suddenly, I felt like I was looking at a picture of dead people. And I sort of was, because none of us were the same people we'd been that day, especially not me.

Below that was a shot of me gleefully opening a gift. Mom and Dad both faced me, but they looked at each other, smiling little secret smiles, the way they used to whenever I did something they enjoyed. With their eyes, they said to each other, *Isn't this great?* Or, *Isn't our girl cute?*—or *funny?*—or *smart? Are you seeing this? Yes. Great. Cute. Funny. Smart.*

But Mom didn't give Keene—or anybody else—those secret, knowing looks now. Keene surely didn't give them to Mom. Dad and Suzanne didn't share happy glances either—at least, not over me—maybe over Baby Robert. Apparently leftovers just aren't that great or cute or funny or smart.

I felt something wet on my shirt. It was only then that I realized I was crying. For that unsuspecting little girl in the snowflake pajamas. I felt as though I were watching her have the time of her life, in the middle of the street, with a big truck speeding straight at her. I wanted to warn her. And then I didn't want to warn her. No, let her have her last bit of pure happiness, because there was no stopping that truck—it would hit her no matter what. And then she'd be gone.

I put the albums away, crawled back into bed, and cried myself to sleep. *I'm just tired,* I thought over and over, and I promised myself I'd be better—stronger—when I woke up.

But when I opened my eyes, nothing had changed, except that my room was dark and Genghis glowed an angry, red 8:42. I didn't feel any better. I felt exactly the same. I couldn't remember ever feeling worse than this. It was either cry—some more—or cook.

I tiptoed to the bathroom, washed my face with cold water, wiped the counter and sink, and then folded the towel and hung it back up exactly the way it had been. Then I went downstairs and waited for a commercial. When one came, I plastered a smile on my face and said, "If you'll let me cook, I'll make anything you want."

Mom looked at her watch.

Keene said, "*Anything?*"

Mom turned and gave him a look.

"What?" Keene said. "She said 'anything.' And I've been craving pineapple upside-down cake all day—we usually have that at family reunions."

Before Mom could respond, I said, "One pineapple upside-down cake, coming right up!" and hurried into the kitchen.

After I'd made the caramel, as I whisked flour, almonds, and baking powder, I felt myself relax, just a little.

Keene took one bite of the cake still warm from the oven and said, "*Oh*. Fizzy, you're like an artist in the kitchen. This cake is a work of art. For the mouth."

I smiled a real smile.

"Thank you," Keene said like he meant it, like he was really glad that my cake and I were there.

Note to self: More cakes and less mistakes.

Chapter 31

Over the next two weeks, I received a few more phone calls from Aunt Liz. I'd almost broken down and called her back, too, because I was in a weakened state—on the verge of tired-tears *all the time*—being perfect is exhausting work.

I'd been getting up twenty minutes earlier every morning to make my bed and straighten my room before school. After school, I made cake. At night, I spent extra time on my homework and studied harder for tests. Plus, I cleaned the tub after I used it and tried to make the bathroom look like I'd never been there, because surely Keene wouldn't mind having a guest he hardly noticed—a guest who made good cakes.

But as tough as things were at home, they were a little better at school.

Mrs. Ludwig had been wearing pants more often, so I wasn't as distracted by the hole in her leg. My math grades improved, although I still made the occasional B.

But today, I was having trouble concentrating on my math because Mara kept tapping me. I made a huffy sound and finally turned around.

"Look," Mara said, pushing back the hair around her ears. "I got new earrings."

"Nice," I said. And then I turned back around and continued working—at least, I tried.

Tap. Tap. Tap. Tap. Tap . . . Tap! Tap! TaaaaaAAAAP!!!

I turned and gave Mara the tired look I so often get.

"They were *very* expensive," Mara said.

"What?"

"My earrings."

I nodded and went back to work.

Tappity-tap-tap-tap!

I ignored Mara.

"Psssst! (*tap, tap*) Pssst! Fizzy!"

"Mara Tierney, are you ill?" Mrs. Ludwig asked.

I kept my eyes glued to my paper, but I heard Mara say, "No, ma'am."

"Are you injured?" Mrs. Ludwig said.

"No, ma'am."

"Then what is so urgent?" Mrs. Ludwig stood from her desk. "What is *always* so urgent? Can't you see that Fizzy is trying to work?"

Silence.

Mrs. Ludwig crossed her arms and stared at Mara while she waited for an answer.

This went on for so long that I had time to go from being irritated at Mara and grateful to Mrs. Ludwig, to feeling sorry for Mara and wishing Mrs. Ludwig would just sit down—the tension kept growing and growing, until I was scared to move a muscle or even breathe too deeply.

Finally, Mara said in a barely audible voice, "Well . . . it's just that I got new earrings."

"I see," Mrs. Ludwig said calmly. "So you're having an emergency *earring* situation. Are you in pain? Do you need help getting them off?"

"No!" Mara said.

Mrs. Ludwig uncrossed her arms. "Then you—and your earrings—will move to the back table, where you'll sit from now on. Do you understand, Mara?"

"Yes, ma'am."

No more tapping! Ever! I thought happily, and then I went back to feeling grateful toward Mrs. Ludwig. I mean, she was sort of protecting me— Hey! Maybe she was even starting to like me!

That afternoon, as soon as we reached Zach's house, his grandmother appeared in the doorway and stayed there. It was like she *wanted* us to know she was watching. Weird.

"Is she mad?" I whispered.

"Nah, that's just her face," Zach said.

"Seems like she'd be happy that you didn't have to stay after school today," Miyoko said.

Zach laughed. "She told me I'd better not have to, because there's a lot of work to be done here. She probably thinks I took too long getting home—I'll probably get the speech on dillydallying and durtling and whatnot as soon as I set foot inside."

"*Durtling?*" Miyoko said, giggling. Apparently that word struck her as hilarious, because what started as a giggle turned into uncontrollable laughter.

Zach grinned. "She means *dawdling*, but she says *durtling*."

"Does 'the speech' usually start with a smack on the head?" I asked Zach.

Zach laughed. "Only when Gran's gotten a call from the school about my behavior while I was there."

The screen door squeaked open and Zach's small grandmother stepped out onto the porch.

"Later," Zach said, jogging toward the porch.

He clomped up the steps and planted a kiss on his gran's cheek.

She said something to Zach.

Zach responded by dumping his backpack, lifting his grandma off her feet, and twirling her around the porch.

I heard Gran say, "Stop it!" But she was smiling ear to ear. They both were.

I smiled, too.

I was still smiling as I started for the kitchen at home when I realized I'd somehow tracked mud into the house. I checked the bottom of my shoes: Sure enough, my left heel had a big clump of mud stuck to it, which had left a trail from the front door into the bathroom and back out. I sighed, slipped off my shoes, and placed them in the bathroom sink—for rinsing—later.

I cleaned the mud up with wet paper towels until there was absolutely no trace of it, and then stood to admire my work. *Oh no*, I thought as I looked over the wooden floor. Now parts of the floor were too clean, which only highlighted how dull and dirty the rest of the floor was. I started to leave it—after all, I had a red velvet cake to make—but then I remembered how important floors are to Keene. *Ridiculous!* I thought. I imagined Keene on his deathbed, saying in a raspy voice, "Remember what's really important in life . . . *floors*"—so I mopped the foyer, bathroom, dining room, and kitchen, too.

I didn't realize just how tired my arms and shoulders were until I started making my cream cheese frosting. I'd forgotten to get the cream cheese and butter out of the refrigerator to let them warm and soften before I tried to beat them together with the vanilla extract and powdered sugar. I made up for my mistake with extra beating time—*lots* of extra beating time.

Keene was the first to arrive home. The minute he stepped inside, he announced, "Wow. It smells like heaven in here."

I peered out of the kitchen and offered him a smile.

"Are you making lemon cake?" Keene asked.

"No, sir, red velvet," I said, confused.

Keene nodded.

I stepped back into the kitchen and then realized that Keene was referring to the lemon-scented floor polish. *He would hope that heaven smells like floor polish*, I thought. *Mom should probably be dabbing that stuff behind her ears, instead of her flowery perfume.*

Luckily, Mom didn't notice the lemony smell when she came home—so I didn't have to tell her how I'd tracked mud into the house. She came rushing through the front door, saying, "I know, I know: I'm running late! I needed to get my chicken pot pie in the oven an hour ago! I'm hurrying as fast as I can!" And she was.

But she calmed down once dinner was on the table. By the time we'd finished, and Mom and Keene were eating my cake—and raving about it—I was so sleepy, all I could think about was how badly I wanted to put my head down on the dining room table and close my eyes, just for a few minutes. But I still had homework to do, a bath to take, and then a bathroom to clean. For the rest of my foreseeable life. *Ugh*. That thought made me want my bed. I thought about how good it would feel to slip between the sheets, lie down under the ceiling fan, close my eyes, and float away. But I knew I couldn't. Not yet. So, I said thank you, forced myself up from the table, up the stairs, and got on with it.

Chapter 32

Mom was the first one to leave the house on Friday morning because she had an important breakfast meeting downtown. She couldn't be late, she told Keene and me so many times that I was tempted to ask, *You don't think we* make *you late, do you?* But I didn't want to start a conversation that might make Mom late.

As usual, once she was gone, the house felt tense and uncomfortable, so I tried to be especially fast to get out of there. But I couldn't find my shoes. I tiptoed around, looking everywhere. I'd just opened the cabinet underneath the kitchen sink, where Mom had hidden my shoes once before, when Keene cornered me.

"If you're looking for the shoes you wore yesterday, they're gone," he said.

"Gone?" I repeated.

Keene nodded. "From now on, any shoes that aren't put away are mine."

"You took my shoes?"

"No, I found them—finders keepers," he said casually, like he was just sharing the weather report.

"I'm sorry... I forgot... I... I was tired and... I just forgot," I stammered. It was only then that I remembered leaving my shoes in the bathroom sink.

Keene nodded like he understood, but he didn't move to get my shoes. "You wouldn't want to wear them anyway—they're dirty." He made a face.

Since I don't have a Lush Valley *collection* of shoes, but only a few pairs, losing one was a big deal to me. Losing a pair to Keene—for the sake of sheer meanness!—was an even bigger deal to me. But I decided to take it up with Mom later.

I said nothing more to Keene, just trudged back upstairs and put on my old moccasins. But before I left my room, I stopped to add to Keene's Dislike List:

18) *Steals shoes—that don't even fit him!—just to be mean!!!*

That made me feel a little better.

Keene was pouring coffee into a travel mug when I came back downstairs. "Do you need a ride?" he asked.

"Nope," I said, and then I was out the door, thinking, *Take that! No "sir" and no ride!* I thought I was punishing Keene, but it didn't take me long to figure out that I was the one I'd actually punished: I mean, I was the one who had to walk, and I was the one who was now going to be late. To school. Again. *Note to self: Punishing others by not allowing them to help you isn't a good punishment—for them.*

I hurried past Zach's house, knowing he was long gone. And

all the while, I wondered if Mr. Moss would let me into science class or send me down to Mrs. Warsaw's office for a tardy slip. I figured the latter was more likely. Then I had an idea.

As soon as I got to school, I went straight to Mrs. Sloan's office, paused, and knocked on the open door. When she looked up, I said, "How's Judas?"

Mrs. Sloan's face went from smiling to confused, and then a light went on behind her eyes. "Oh yes, my cat. He's fine, Fizzy. Thank you," she said. "Do you need a tardy slip this morning?"

"Yes, ma'am," I admitted sheepishly.

Mrs. Sloan nodded. "Well, come in and sit down for a minute while I look for my pen and pad."

I sighed, stepped into her office, and closed the door behind me. I dumped my backpack on the floor and dropped into a chair at the little worktable.

Mrs. Sloan got up and moved books and piles and files around on her messy desk until she found her pink tardy slip pad. She held it up as if to say, *Ta-da!*

I nodded and said, "Um, there's a pen in your hair."

Mrs. Sloan felt around in there and wrestled the pen from her curly hair with an aha! She brought the pen and pad with her when she sat down at the table with me, but she didn't do anything with them—like write me a tardy slip. Instead, she set them off to the side.

I knew the questions were coming, so before she could ask any, I asked one of my own: "Do you ever have overnight guests?"

Mrs. Sloan laced her hands together on the table. "Yes."

"Do you ever wish they'd go home? Like maybe you don't want to make your bed? Or maybe you want to leave your dishes in the sink and stay in your pajamas all day? Or maybe you want to kick off your shoes and leave them right by the front door?"

"Sure."

"Well, that's how it is when your parents get remarried," I informed her. "You want your stepparents to go home after a while, or you want to go home, but nobody ever gets to go home again."

Mrs. Sloan didn't react, didn't say a word, and didn't move a muscle, but something in her eyes hardened.

I thought maybe she didn't understand. "It's sort of like you've adopted a guest, because you have to be on your very best behavior . . . forever. But actually, *you* are the guest."

Mrs. Sloan unlaced her hands, placed them on the table, and leaned forward just a little. "What makes you the guest, Fizzy?"

"I'm a kid," I said, shrugging one shoulder. "I don't have a job or a house of my own. So I'm counting on somebody else to let me stay in their house, somebody who doesn't *have* to let me stay if they don't want to."

"And by 'somebody,' you mean your stepparents," Mrs. Sloan said.

"Right."

"What about your parents?" Mrs. Sloan asked. "Have you tried discussing this with them?"

I gave her a look like that was just about the dumbest idea I'd ever heard. "*No, ma'am.*"

"Why not?"

"Because that wouldn't be polite—it would probably hurt their feelings. My parents *chose* my stepparents—they love them."

Mrs. Sloan leaned back in her chair and sighed. "Don't you think your parents love you, too?"

"Yes, ma'am."

"If they care about your feelings anywhere near as much as you care about theirs, then I think you should talk to them. There are more important things in life than manners, Fizzy."

I stared at her. "You're not from around here, are you?"

Mrs. Sloan laughed and reached for her pen and pad. She wrote me a tardy slip, but before she handed it to me, she looked me deep in the eyes and said, "I'm so glad that we're becoming friends, Fizzy."

As soon as she saw me through the little window, Miyoko jumped up from her desk and ran to open the door, before Mr. Moss even knew what was happening.

"Thanks," I whispered.

Miyoko smiled and we both hurried to our seats, me tossing my tardy slip on Mr. Moss's desk on the way.

Meanwhile, Mr. Moss continued with his lesson, barely casting a glance in my direction.

Safe, I thought, and I let my guard down a little.

That was a mistake, because when the lesson was over,

before he sat down at his desk, Mr. Moss turned to glower at me as he said, "Miss Russo, there are exactly eight days of school remaining, and I'd appreciate it if you were on time for every single one of them."

I felt my cheeks heat up and tired-tears spring to my eyes as I nodded. *Don't cry,* I told myself. *The worst is over. The day can only get better from here.*

But I was wrong. That afternoon, my locker jammed between gym and math class—and I couldn't go to math without my math book. I just couldn't. So I had to go down to the office, where they paged the school janitor. By the time I finally got my book and made it to Mrs. Ludwig's room, class was halfway over. The lesson was finished and everybody, including Mrs. Ludwig, was working quietly at their desks.

Both Miyoko and Zach looked up and smiled when they saw me.

I smiled back.

But Mrs. Ludwig didn't smile. She gave me a disapproving look over the top of her glasses and held out one hand.

I gave her my—*excused*—tardy slip.

She looked it over, unimpressed, set it aside, wrote on a notepad, *Page 265, section A, problems 1–30,* tore the paper free, and handed it to me, all without uttering a single word.

I settled at my desk, turned to page 265 in my math book... and had no idea how to do the problems on that page. I flipped backward, read the lesson, and stared at the example, trying to figure it out. But I couldn't.

Then I looked around the room, trying to tell if anybody

else was having trouble, but everyone was working steadily. No one looked lost. Like me.

"Keep your eyes on your own paper, Fizzy Russo," Mrs. Ludwig said.

My heart kicked and began to race. "Oh no . . . I wasn't . . . I . . . just don't know how to do these problems."

"Is that so?" Mrs. Ludwig said sarcastically, like *What a shocker.*

I gulped and nodded.

"Then perhaps you'll be on time for our lesson tomorrow."

"Yes, ma'am," I said.

Mrs. Ludwig went back to grading papers.

Nope, she still doesn't like me, I thought, feeling disappointed somehow.

Christine Cash took her paper up to Mrs. Ludwig and they did some whispering. Then Christine went back to her desk, got her chair, and moved it next to Mrs. Ludwig's at her desk, where they continued working together.

I wanted to move my chair up there, too, and have Mrs. Ludwig whisper math secrets to me. But since I obviously wasn't welcome, I stayed where I was, staring at my math book, willing understanding to jump off the pages and into my brain. But it didn't. So, after that, I just sat there trying not to cry, because getting a B on an assignment is one thing, but an F? I mean, at least a B says, *Hey, I tried.* Whereas an F is so far from perfect, it pretty much says, *Who cares? Not me.* How could I defend an F? Just the thought of trying made me feel sick.

When the bell rang, I got out of there as fast as I could. When I neared Mrs. Sloan's office, she came farther out into the hall to meet me, putting her warm, plump arm around me, and saying, "Come on, Fizzy. Come with me."

I shook my head and whispered, "That's okay. I'm fine."

"Please. Come," Mrs. Sloan insisted.

I slumped into my usual chair at the worktable as Mrs. Sloan closed the door. "It looks like you might be having a tough day."

"I'm okay," I said, even as my chin quivered and tears rushed to my eyes.

"Did something happen in Mrs. Ludwig's room?" Mrs. Sloan guessed.

My head snapped up. "Why? Did she say something about me?"

Mrs. Sloan smiled. "No, but I just saw you come from her room."

"It's not Mrs. Ludwig," I said as Mrs. Sloan sat down with me. "I mean, she's not helping, but she's not the real problem."

"What's 'the real problem'?"

I didn't answer.

After a minute or so, Mrs. Sloan nodded as if I'd spoken. Then she said, "I've given your words this morning a lot of thought, Fizzy, and I want you to know I understand."

"You couldn't," I said, "not really."

Mrs. Sloan folded her hands in her lap, took a deep breath, and said, "My mother died when I was a young girl and my father remarried rather quickly. It was ... difficult."

I was so relieved to hear this, to know that someone—anyone—understood, that the tears spilled out over my eyelashes and down my cheeks. But I ignored them and asked, "Did it ever get easier?"

"Not for a long time," Mrs. Sloan answered honestly, "but yes, it did get easier. And better."

I wiped my cheeks with my hands and then wiped my hands on my jeans. "When?"

"For me, things got easier slowly as I came to know my stepmother, as I learned what was important to her and what wasn't, what she wanted from me and what she didn't. But even then, things were still often tense between us. Until I moved out of the house."

"What happened then?"

Mrs. Sloan smiled. "We missed each other. I think we were both surprised by this. I know *I* was surprised. And then I realized that at some point, when I was no longer expecting it or hoping for it, my stepmother and I had become family."

"That's never going to happen with my stepfather and me."

"Maybe not. But just because you feel that way doesn't mean it won't happen—I never thought it would happen either."

"You don't understand," I said. "He doesn't *want* me."

"Do you want him?"

I lowered my eyes. "No, ma'am."

Mrs. Sloan was quiet for such a long time that I had to look up at her to make sure she was still awake.

She was. "I like your necklace," she said.

I touched the tiny cross in the hollow of my neck. "Thank you."

Mrs. Sloan leaned over the table like she was about to tell me a secret. "My stepmother and I didn't want each other either. But, thankfully, sometimes God blesses us with people we never asked for or wanted—because He knows we need them, even if we don't."

"I don't need my stepfather," I insisted.

"I didn't think I needed my stepmother either, but I did, Fizzy. In so many ways. And even if you really don't need your stepfather now, that doesn't mean you won't need him later."

"I won't," I said. "I really won't."

"What if your children need him? Fizzy, my stepmother was the only grandmother my children ever had. They adored her. And she adored them because they were the only grandchildren she ever had."

"But he took my shoes!" was all I could think to say.

Chapter 33

The first thing I saw when I walked in the door at Dad's was a huge portrait of the four of us hanging above the fireplace. It wasn't a great picture of me, but I loved it anyway. I loved it because the four of us, in that picture, in that moment looked . . . like a family.

"Like it?" Dad asked.

"I love it," I said.

"I love it, too," Suzanne said, coming up behind us with Baby Robert in her arms.

We all stood looking at the portrait and I noticed we felt like family. I felt like family, standing there with them. I remembered what Mrs. Sloan had said then and thought maybe she was right. Maybe it was happening. Maybe we were becoming a family!

I was happy for the rest of the night. Happy even though meat loaf showed up on the dinner table, happy even though I came in dead last at a game of Scrabble—I was distracted by Baby Robert's cuteness. Just *happy*.

It didn't even bother me when Suzanne brought the pictures of me alone to my room to show me. They were good pictures—except for all the freckles—and I knew Mom would love them.

• • •

I was in bed watching my little TV when Baby Robert started crying. I turned the volume up and watched two more shows. That's how I knew that Baby Robert had been screaming for more than an hour, and that's when I started to think maybe I could help—I could try singing Miyoko's lullaby, "Nenneko yo."

I got out of bed and followed the dreadful sound down the hallway to Dad and Suzanne's bedroom. The door was closed, so I knocked and waited. Nothing. I thought maybe they couldn't hear me over all the wailing, so I knocked again, louder.

The bedroom door flew open and Suzanne stood before me with Baby Robert in her arms. She didn't exactly look happy to see me.

"What?" Suzanne barked over Baby Robert's yowling.

"I heard the baby . . . ," I said, looking past her, hoping Dad was in there and that he'd come to my rescue. But I didn't see him. What I *did* see was another picture, a little smaller than the one over the fireplace, only this one had just the three of them in it. The homesickness spread through my belly like an egg that had been cracked open.

"*What?*" Suzanne repeated.

I swallowed. "Nothing."

As I started to turn away, the door practically closed in my face.

I found Dad downstairs in the kitchen, warming up a baby bottle.

"I'm calling Mom," I announced. "I want to go home. I'm *going* home."

I think Dad would've looked exactly the same if I'd hauled off and kicked him in the shin—surprised, confused, and very unhappy.

I picked up the phone.

"Wait," Dad said. "What happ—"

"Robert!" Suzanne yelled from upstairs.

"Coming!" Dad answered, testing the bottle on his wrist. "Wait," he said to me on his way out of the kitchen. "*Just wait.*"

Mom didn't answer the phone at home, so I called her cell.

By the time Dad came back downstairs looking for me, I was dressed, packed, and watching out the front window for Mom's car. A flash of lightning lit up the night sky.

"Fizzy," Dad started just as Keene's car pulled to the curb. Why did that car always make me feel so *disappointed*—and nervous?

Thunder rumbled in the distance. "Mom's here," I said, avoiding Dad's eyes. "I've got to go."

Before Dad could say anything more, I was out the door and running as fast as my legs—and suitcase—would go. Wind rustled through the trees as the storm approached.

Keene sat in the driver's seat. Mom was beside him. They were both dressed up, like maybe they'd been somewhere fancy when I called. Before I even got in the car, I could tell that Keene was mad.

He didn't want to come and get me, I realized. For a second, I thought about going back inside Dad's house, but then I knew Dad and Suzanne didn't want me either.

I wondered, *What does it say about you when even your own family doesn't want you anymore?* I felt sure it indicated that there's something seriously, severely wrong with you. With *me*.

Even though I was trying my hardest to be perfect.

But since they'd stopped what they were doing and come all this way, and since big raindrops were starting to splatter down on me, I went ahead and got in the car with Keene and Mom.

Mom turned around in her seat and said softly, "Fizzy, honey," and that was all it took.

I burst into tears. I don't know why Mom has this effect on me, but she does. Once, when I was eight, I wrecked my bike near Olivia's house, skinning my thigh. But I didn't cry. Instead, I hopped right up and told everyone I was fine. When Olivia's mom said that my leg was in bad shape and she needed to call my parents, I was a little scared, but still, I didn't cry. When Dad showed up to get me, I didn't cry then either. But as soon as I walked through the back door and saw my mom, I fell to pieces. "She was fine a minute ago," Dad kept saying, like I was faking or something. I barely had any skin left on my thigh; I wasn't faking.

Anyway, once I started crying, I couldn't stop. I cried so hard and loud that eventually Mom stopped trying to talk to me, and instead just reached into the backseat and put her hand on my knee. Rain began pounding the windshield. I did that kind of crying where you make weird sounds that you never make otherwise and you can barely breathe.

By the time we got home, I could tell the anger had sort of melted off Keene. Now he just looked . . . uncomfortable.

It didn't make any difference to me. I kept right on blubbering as Mom hurried me into the house. My teeth chattered violently. I felt cold and wet right down to my bones. My body felt heavy, my head pounded, and my stomach sloshed around like an out-of-control Tilt-A-Whirl. I felt more homesick—for a place, a time, a family that didn't exist anymore—than I ever had in my life.

Mom took me upstairs, helped me out of my rain-soaked clothes and into my pajamas, tucked me into bed, and sat down beside me. Then she brushed my hair with her fingertips and said over and over, "It's going to be all right," until I started to calm down.

But even after I stopped crying, I couldn't get hold of myself. My breathing was still funny—like hiccups or something, only not hiccups—and I kept shuddering.

"You're awfully warm," Mom said then. "Do you feel all right?"

I opened my mouth—to say that I was fine—and promptly vomited all over my bed.

"Keene!" Mom shouted. "Keene!"

I blindly reached out, caught Mom's arm, and squeezed as I continued retching. I wanted to say, *No, don't let him see me like this!* Only I couldn't say anything at the moment.

Mom must've misunderstood the arm squeeze, because she responded by pulling my hair back and holding it.

I saw Keene's polished, black-tassel loafers step into my

room. But as soon as he'd had a few seconds to take in the scene, his shoes turned and carried him away—quick. I didn't blame them—or him.

"Keene!" Mom called sharply.

Keene's shoes reappeared in the doorway.

I raised up and wiped my mouth on my hand.

Mom let go of my hair and placed a strong arm around me. Then she said to Keene, "I need you to strip this bed while I get her into the shower."

Keene didn't look like he wanted to, but he didn't argue.

Mom was sitting on the toilet lid in the bathroom, waiting for me, when I stepped out of the shower. She handed me a towel and said, "I've put fresh sheets and another blanket on your bed."

"Thank you. I . . . I'm sorry," I said.

"It's all right," Mom assured me.

When I was back in bed, she said, "Now then. We'll talk in the morning—it'll all look better in the morning, you'll see."

I nodded obediently. I knew Mom was wrong, but she was trying so hard.

"Sleep now," Mom said. "Just sleep."

I was listening to the rain, about to drift off to sleep, when I heard the phone ring. Somehow, I knew it was Dad calling. I felt like I should get up and tell him how sorry I was—for everything—but I was too tired to move. Mom's words echoed in my throbby head: *Sleep now. Just sleep.*

Chapter 34

I felt so far from perfect and so ashamed the next morning that I didn't want to get out of bed. And I definitely didn't want to come out of my room. I didn't know how I'd ever be able to come out and face anyone again.

But I didn't have a choice, because eventually Mom came in, bringing a tray of food with her. I sat up and Mom placed the tray on my lap: green tea, beef broth, toast, orange Jell-O, a napkin, and two spoons—because I don't like my silverware to be cross-contaminated with food other than the one I'm using it to eat.

"Thank you," I said, and I meant it, because even though I wasn't hungry, somehow that tray made me feel like I was better than leftovers, like I *mattered*, like someone cared how I felt—even about silverware.

"You're welcome," Mom said, pulling the chair from my desk over to sit down beside me. "How're you feeling?"

"Better," I said.

"So . . . ," Mom said, and I knew she was waiting for me to explain last night.

"I was sick," I said, hoping this would be enough.

Mom waited for me to say more.

"With sickness," I added. "I'm sorry."

"That's all right," Mom said easily. "Everybody wants to go home to their mom when they aren't well. But, Fizzy, are you sure that's all it was? There's nothing else bothering you?"

I thought about this. What was bothering me most this morning was that, when only Mom and I lived in the town house, I'd looked back on the times when we lived at home with Dad as the good times. Now that Keene was here, I looked back on the times when it was just Mom and me in the town house as good times, too—good times I didn't even appreciate until things were so much worse. And then I wondered if at some point I'd look back on *this* time and think the same thing. I figured things could only get worse, seeing as how there had to be something seriously wrong with me. But I couldn't tell Mom any of that. My misery would take a great big old bite out of her happiness. I knew it would. I just knew.

Apparently I took too long deciding this, though, because Mom said, "Tell me what you're thinking about."

I said what I hoped would amount to a tiny crumb instead of a whole bite: "Did you know that Keene took my shoes?"

"'Took your shoes'?" Mom said. "What do you mean?"

"I left them in the bathroom because they were muddy and I planned to clean them after I mopped the floors, but I forgot and he took them. He said, 'Finders keepers.'"

I could tell that Mom didn't understand and, more important, didn't approve. But all she said was, "I'll get your shoes for you—and I'll talk to him. What else is on your mind?"

I searched my mind for something else smallish and

safe-ish—crumblike—and finally came up with, "I don't like my math teacher."

"All right," Mom said easily. "Why not?"

"Because *she* doesn't like *me*."

Mom started to smile but she caught herself and said, "How do you know? Does she treat you differently than the other kids?"

"Maybe not," I decided. "You're probably right: She doesn't like any of us."

Mom did smile then. "Just remember: There's a difference between being mean and being tough. Tough teachers are usually good teachers, but you'll have a new one soon enough—you're almost a seventh grader now."

That reminded me: "Don't you think a seventh grader ought to have her own phone?"

Mom shook her head. "Fizzy, the more you ask about the phone, the more you reveal your failure to accept that you can't always get everything you want in life."

That meant two things: 1) Mom didn't want to hear another word about the phone, and 2) if I kept on bugging her, I'd *never* get a phone. *Elle est diabolique, non?* ("She is diabolical, no?") I nodded and picked up a spoon, not because I planned to eat, but because I was finished talking.

I called Dad later that afternoon and told him I was sorry.

"What happened?" he said.

"I don't know. I felt tired and sick and . . . I just had a little meltdown, I guess. Can you forgive me?"

"You're forgiven, Fizzy, but next time, let's talk things over, okay? Running away never solves anything. And we know how to take care of sick kids here, too."

"Yes, sir," I said.

"Should we try it again next weekend?"

"Next weekend would be good," I said.

Since that had gone okay, I called Aunt Liz, too.

"Hi! How are you? Oh, Fizzy, I miss you so much!" Aunt Liz gushed, not sounding inconvenienced or irritated in the least.

I exhaled—I hadn't known I was holding my breath until then. "I'm fine, just really busy," I said. *Being perfect*, I thought. "How are you?"

"Good—this morning, I made a flourless, sugarless chocolate cake that looks promising."

"Why?" I thought out loud. "Why would you bother making a cake without the best parts?"

Aunt Liz laughed and then we spent a few more minutes talking cake.

"Speaking of cooking, have you heard from *Southern Living*?" she asked.

"Not yet," I said, "but I should any day now."

After I put the phone back, I heard Mom calling for Keene in a frustrated voice. Naturally, I went to see what that was about.

At the bottom of the stairs, I heard Mom say to Keene, "I can't find Fizzy's bedclothes anywhere—I need to wash them."

Keene shook his head, like that was just about the craziest thing he'd ever heard, and said, "I threw all that stuff away."

"*What?* Where?" Mom said.

"Outside," Keene said, "in the Dumpster."

"Go get it, please," Mom said with forced patience.

Keene shook his head again. "That stuff's nasty, Cecily—you just don't—"

"I'll get it," I said, interrupting him. I mean, they were *my* bedclothes and it was *my* puke on them.

Keene turned and gave me a look that I interpreted as, *Thank you.*

I nodded my head once.

I was about to close the door behind me when Keene said, "Hose that stuff off outside before you bring it into the house."

I rolled my eyes—after I shut the door.

When I returned to my room, I could hear Mom and Keene talking downstairs—through the vent:

"We all have our little pet peeves, Keene—even me," Mom said, "but can you name one of mine?"

"No," Keene said.

I lay down on my belly and put my ear close to the vent.

"That's right," Mom said, "because I think of them as *petty* peeves. But if you really want to know, it bothers me when you leave your briefcase, car keys, mail, and sunglasses on the dining room table."

"That's where I've always put them," Keene said. "I'm sorry—I didn't know."

"Like Fizzy didn't know your pet peeve about floors until recently. It takes time to adopt new habits. Give her time. Remind her if you have to, but please don't take her things."

Keene didn't say anything.

"You know," Mom said, "once, I thought about picking up all your stuff off the dining room table and dumping it in the passenger seat of your car—I know how particular you are about your car. But then I thought about how lucky I am to have you, *and your things*, here with me."

Keene still didn't say anything. What could he say? It's not like he could say he was glad that me and my things were here—because he wasn't, I knew. Mom must've known, too, because she didn't say anything else either.

That evening, after dinner, Mom said, "Keene, please give Fizzy her shoes back now." So Keene did, but he didn't apologize for having taken them in the first place, which is why I didn't apologize for having left them out.

"From now on, if you leave your shoes out, we'll remind you to put them away. If you don't . . . well, you'll get fair warning before you lose your shoes," Mom said, but she wasn't looking at me; she was looking at Keene.

He looked a little pouty.

I pouted, too—because no one should take my shoes, not even with a warning! They're *my* shoes! MINE! I mean, do we teach thieves that it's okay to steal *if* they give their victims a warning first? I did some huffing and then took my shoes up to my room.

But I tried to make up for everything by leaving the bathroom extra sparkly that night. I scrubbed the tub and all the fixtures until they were gleaming. Then I went ahead and did the counters and two sinks, too—even cleaning in between the handles and the faucet with an old toothbrush. I thought about going downstairs to get the mop but didn't want to call attention to myself—or interrupt anything—so I ended up cleaning the floor with some old washcloths.

I was still thinking about the importance of floors when I went to my room. I noticed right away that it needed vacuuming. So I set the—hateful—Genghis extra early, knowing he would enjoy stealing a little more of my sleep. But when I turned out the light and slipped into bed, I found the truth waiting for me:

No amount of cleaning or cooking or studying would change the fact that there was something wrong with me. My family knew it. Mrs. Ludwig knew it. Mrs. Warsaw knew it. Even my former best friend, Olivia Moore, probably knew it. And now I knew it, too. What I *didn't* know was what it was or how to fix it.

But, I told myself, *maybe winning the* Southern Living Cook-Off *would fix it*. If not . . . well, I figured it would at least justify the space I took up on the planet, the oxygen I used, the food I ate. Yeah, if I won, I'd probably be forgiven for . . . *existing* . . . right? Surely I would. I really needed to win that contest.

Chapter 35

Two envelopes were practically burning holes in my shorts pocket as Miyoko and I walked home from the last day of school. But I only mentioned one of them to Miyoko.

"Did Mrs. Ludwig give you an envelope on your way out of her room today?" I asked.

"No," Miyoko said. "Why? Did she give you one?"

I nodded. "Zach got one, too—I was behind him in line, so I saw—and if Zach and I are the only kids who got one . . . well, you *know* it's bad."

Miyoko didn't seem to disagree. "Let's just open it like before—at least then you'll know what you're up against."

I wasn't sure.

"I'll throw the envelope away at my house and you can just pretend there wasn't one."

"But what if it says something like, 'If the seal on this envelope has been broken, the letter has been tampered with,' that'd be very coplike, don't you think?" I worried.

Miyoko laughed. "It doesn't say that."

I gave her a doubtful look.

"It doesn't," she insisted.

I decided Miyoko was probably right, pulled the envelope

out of my pocket and unfolded it as we huddled up on the sidewalk to look at it.

The outside of the envelope read *Fizzy Russo*, not *To the Parents of Fizzy Russo*. It was for me. Huh. I sort of wished I'd looked at it sooner, but I made up for it now and ripped the envelope open:

Dear Fizzy,

As you probably realize, I have pushed hard and been tough, especially on you. What you may not realize is that I did this because nothing upsets me more than wasted potential. I knew you were an A student, making Bs when you should've been making As. So, I pushed for more, and you gave it. As a result, you have earned an A in math this semester.

I expect you to earn even more As next year, and have told your new math teacher so. Therefore, you can expect him to be tough on you, too. But know this: As long as he's being tough on you, he believes in you. It's when a teacher stops being tough on you, stops pushing you, that you should worry—because that's when they've given up on you. But no one is going to give up on you, Fizzy, least of all me.

With great expectations,
Mrs. Ludwig

When I finished reading the letter, I had tears in my eyes, but I blinked them back quickly and said, "I've never had a math teacher who was a man before."

Miyoko smiled knowingly, but didn't comment on my tears or try to hug me or do anything to indicate that she'd noticed them—she is an *excellent* friend.

I stuffed Mrs. Ludwig's letter back into my pocket and left the other envelope where it was. I didn't want to tell Miyoko about that one—which was weird because I'd talked about receiving it almost nonstop until I did. I'd been waiting for it for months.

Then last night, there it was: an envelope with *Southern Living* written across the top left corner, sitting on my dresser. I knew—because of where I found it—that Keene had gotten it out of the mailbox, and I checked to make sure he hadn't opened it and read it. He hadn't. He hadn't even read the outside of the envelope—apparently—because he never said a word about it. Neither did Mom. So, whatever the letter said, it was between *Southern Living* and me. Just the way I wanted it.

But then, for some reason, I couldn't bring myself to open it. I mean, what if the letter said my recipes stunk? What if it said something like: *Dear Miss Russo, We regret to inform you that not only have your recipes not qualified for the cook-off, they've made us sick. So we here at* Southern Living *magazine would like to take this opportunity to suggest that you try something other than cooking, Miss Russo . . . anything other than cooking.* What then?

I wasn't ready to give up on my dreams yet. I wasn't ready to give up on the idea of winning the *Southern Living* Cook-Off. I wasn't ready to give up on becoming a world-famous chef. And I certainly wasn't ready to give up my television show!

I heard Mom somewhere in the back of my mind: "I can't

give up on my dream of having a family either," she'd said. And I thought I'd understood, but now I *really* understood. I decided I'd try a little harder to be friendly with Keene—for Mom.

I slowed to a stop on the corner of Chrysanthemum Court, where Aunt Liz lives.

Miyoko stopped, too, and turned to me, wearing a puzzled look on her face.

"My room's clean, and my mom needs to grocery shop, so I don't have the ingredients to cook today. Plus, I don't have any homework, so . . . I'm going to see Aunt Liz," I announced.

Miyoko smiled. "Good."

"Do you want to come with me?" I asked, hoping—just this once—that she'd say no.

"Can't," Miyoko said, suddenly looking grim. "Tiger mom's waiting."

"Another day, then—maybe tomorrow, since we don't have school," I said, hoping to cheer her up.

"Sure." Miyoko nodded. "Fizzy, is everything okay?"

"Oh yeah," I said, bobbing my head up and down. "I'm just tired, I guess. What about you? Is everything okay with you?"

Miyoko looked down at the sidewalk and seemed to be thinking.

I took a step toward her and whispered, "What is it?"

"My mom's mad at me. It's not like this is new. She gets mad at me all the time, but I never get used to it. It always stays with me and bugs me, you know?"

I nodded. "What happened?"

"She was upset that I gave her slippers for her birthday,

because I gave her slippers for Christmas, too—I forgot. She said my gift lacked any real thought or effort, and was therefore lacking in love."

My eyes bulged.

"My mom thinks I don't love her, Fizzy," Miyoko said, dabbing at the outer corner of one eye with her finger.

I wiggled out of my backpack, let it fall to the sidewalk, and put my arms around Miyoko. "Your mom knows you love her—she's just mad. Please, come with me to Aunt Liz's," I said, and I meant it.

"I really can't," she said, holding on to me like her life depended on it.

"Okay," I said. I gave her a few more seconds, then took a step back. "That's okay. You do whatever you need to do." I figured the very least I could do for Miyoko was not add more pressure.

"Thanks," was all Miyoko said. Then she sniffed and added, "Tell Aunt Liz I said hi."

I knocked and pushed through Aunt Liz's front door, letting myself in.

"Here! Here! I'm back here!" an excited voice called from the sunroom.

Aunt Liz met me in the doorway off the kitchen, where she smiled an electric smile that seemed to spark from the tippy-top of her head right down to her little toes. She threw her arms around me and said, "Oh, Fizzy! You're here! I'm so happy to see you! I've missed you!"

This was the exact opposite of the tired what-now? reaction I often got when I walked into a room. It made me feel happy, so happy that I couldn't even remember why I'd ever been upset with Aunt Liz. And I didn't try. Instead, I breathed in the foody scents from the kitchen, the flowery scents from the sunroom, and the fruity scents of Aunt Liz's hair and perfume and let the homey feelings wash over me.

Aunt Liz pulled back from the hug to look at me. "Where've you been? Are you okay? Are *we* okay?"

"Yes. And yes," I said. "Just busy."

Aunt Liz hugged me again and said, "I wish I'd known you were coming! I would've made Benedictine!"

"That's okay," I said. "It doesn't matter." And it really didn't, because right now I didn't need Benedictine; all I needed was Aunt Liz.

Of course, Aunt Liz poured me some sweet tea anyway—which never hurts—and I followed her back out to the sunroom. Aunt Liz settled into the cushy rocking chair that offers the best view of her rose garden. I sat down in the rocker beside hers. "Well? Tell me everything!" Aunt Liz said. "What've you been up to? Oh! Was today your last day of school? How was it?"

"Fine," I said.

Aunt Liz smiled and watched a fat bumblebee buzzing around her pink roses. "How's Miyoko?"

"Um . . . well . . . ," I said, thinking about Miyoko and her tiger mom—who had graduated to monster mom in my opinion. I mean, what kind of person complains about a *present*? Aren't gifts always good?—like cakes?—and chocolate? I began rocking at a furious pace. I wanted to tell Aunt Liz about Miyoko's problem, only I knew I couldn't because it wasn't mine to share or not share—it didn't belong to me.

Aunt Liz turned and looked at me expectantly.

"Miyoko's fine." I stopped rocking, fished the unopened envelope out of my right pocket, and handed it to her.

Aunt Liz stopped rocking, too, unfolded the envelope, and turned it over. "Fizzy! You haven't even opened this!"

"You open it for me," I pleaded, "and if it says really mean things, don't tell me those parts."

Aunt Liz laughed a nervous laugh. Then she tore the envelope open and scanned the paper, reading quickly.

Still, I could hardly stand it.

Finally, she looked up at me with soft, understanding eyes.

My heart dropped into my stomach, my throat tightened, and I could feel tears gathering behind my eyes.

Aunt Liz broke into a big smile. "Russo Lasagna has made it into the final cook-off!"

I bolted up out of my chair and Aunt Liz did, too. We hugged and jumped up and down together and I said over and over again, "Great gravy! I can't believe it!"

When we started to calm down, I looked at Aunt Liz and said once more, "I just can't believe it! Can you?"

She smiled. "I can believe it, Fizzy. I'm not even that surprised."

I thought that was just about the nicest thing anybody had ever said to me.

Aunt Liz grabbed the long spoon from her iced tea glass, turned it upside down, and held it under my mouth like a microphone. "So," she said, "Fizzy Russo, now that you've qualified for the *Southern Living* Cook-Off, what are you going to do?"

"I'm going to Disney World!" I hollered, like people do on TV.

Aunt Liz and I laughed and laughed.

I read the letter twice. Then I read it again:

Dear Miss Russo:

We are pleased to inform you that your recipe for Russo Lasagna as entered in the Southern Living *Cook-Off has qualified you as one of the top finalists in the Family Favorites category, making you eligible to compete in the* Southern Living *Cook-Off, which will take place on July 11, before a live audience, in Charleston, South Carolina . . .*

Chapter 36

The day that Mom and I were leaving for the *Southern Living* Cook-Off in Charleston, everybody was waiting in our town house's small front yard to show their support and see Mom and me off to the airport.

Aunt Liz brought me a Benedictine sandwich in a brown paper sack, in case I got hungry on the plane.

Dad and Suzanne—and Baby Robert—brought a mini fire extinguisher with a big red bow on it, which was, they said, in case I forgot something in the oven during the cook-off. We all laughed. (I left my fire extinguisher at home.)

Keene held flowers, and I might've thought they were for Mom, except that they were purple. I thanked him. And then he hugged me, which made me think, *Hey, maybe he likes me. But then again, probably not. How could he?* I decided not to think about that any more. Today.

Zach showed up and brought me a new—musical—alarm clock.

Miyoko was there, too, with Mrs. Hoshi—who gave me a Japanese cookbook I could read if I got bored on the plane.

Mom was snapping pictures of Miyoko and me when I overheard Mrs. Hoshi, standing off to the side, talking to Keene.

"They're an unlikely pair, aren't they?" Mrs. Hoshi said.

"Are they?" Keene said.

"Yes," Mrs. Hoshi said certainly. "Miyoko's a very serious girl—she leaves for Super-Scholars Camp tomorrow."

I looked over at them: Keene seemed impressed.

Mrs. Hoshi said, "Meanwhile, Fizzy is off to some cooking contest."

Keene's brow furrowed and he turned to stare at Mrs. Hoshi.

She looked at her watch and then fanned her face with her hand.

I imagined myself walking over to them and saying, *You look so nice today, Mrs. Hoshi . . . like, really, especially nice—you must be wearing your Big Booty Judy Bloomers—right?* That made me smile—a real smile—and I heard Mom's camera click, capturing my expression at this thought forever.

Then I heard Keene say, "Miyoko *is* a remarkable young lady."

I glanced back at them and Mrs. Hoshi was nodding in agreement.

"But Fizzy is remarkable, too," Keene added. "If you think otherwise, you're underestimating her, believe me. This isn't just 'some cooking contest.' It's the toughest cooking contest in the country, and Fizzy's worked hard to qualify for it—competing against adults, many of whom are professional chefs."

Out of the corner of my eye, I saw Mrs. Hoshi stand up a little straighter. "Of course," she said.

I smiled at Keene.

He smiled back.

Zach slipped in behind Miyoko and me and tried to photo-bomb Mom's pictures by making funny faces and holding up bunny ears behind our heads. Mom kept snapping away. She snapped pictures of everybody until it was time for us to go.

We hugged everybody again, said our good-byes, and promised to call as soon as we had news. I really hoped it would be *good* news, because even though I appreciated everybody showing up like this, I also felt obligated to do well—for them—since they'd made such a fuss over me. Even bigger than the fuss was the fact that members of both sides of my family had knowingly, voluntarily shown up in the same place at the same time—while smiling!—for me.

Chapter 37

Looking back, I have to admit that I don't remember much about the city of Charleston. I remember Mom saying how beautiful it was and talking about the Spanish moss that hung from the branches of old trees—"like wedding veils," she said—the historical homes, the wide front porches, and the joggling boards, which are like long benches you can bounce on, but I don't remember any of these things myself. My only lasting impression of the city was that it was as hot as a frying pan, at least in July. I started each day as cool, clean, and dry as sugar, but ended it feeling more like caramel—hot, brown, and very, very sticky. (Caramel is just cooked sugar.)

What I remember best is room service and, of course, the *Southern Living* Cook-Off. Now, I love room service. Love it. Love it. Love it. I also love the fancy silver lids that cover the plates of food when they arrive at your door. After Mom and I ate our meals, I'd practice for hours with those silver lids, because I planned to use them on my TV show. So I'd stand in front of the beds and I'd say to my studio audience (the pillows), "And after baking for fifty-five minutes at 350 degrees, voilà! Russo Lasagna!" And then I'd lift a silver lid with real pizzazz, to show the audience my fabulous creation. Yes, I highly recommend silver lids.

I also recommend taking part in the *Southern Living* Cook-Off. The cook-off was staged in a grand ballroom inside our hotel. When Mom and I made our way through the maze of people, tables, equipment, and electrical cords and found the area for cooks competing in the Family Favorites category, we stopped and chatted with a few of them. One of the cooks was the biggest man I'd ever seen, so I knew his cooking was good because—duh!—look at him! We said hello, but Big Boy only nodded and went back to checking his equipment.

Francois was the first cook we officially met, and he was quick to inform Mom that he was an expert in "molecular gastronomy." Then he looked down at me and said, "That's food science," as if I didn't know what molecular gastronomy meant! I decided I didn't like him—and also that he used way too much hair gel.

The second cook we met was a woman who thought Mom was the one competing against her. When she learned that I was in the contest, she looked me over and said, "Brilliant marketing strategy—a child—what's next, baby zoo animals?" I didn't like her either.

But the lady who would be cooking directly across from me, Ms. Marla of Farmville, North Carolina, was really nice. She wore a blue flowery dress and looked scared—like me—which made me want to hug her. But she came out from behind her station to hug me first.

"Thank you," Ms. Marla said when I hugged her back. "I needed that."

"Me too," I said.

We smiled and wished each other luck.

Then it was time to cook. In front of my station stood a huge man in a black suit with a curly cord running up his neck, from under his collar, into his ear. His hair was dark, his eyes were dark, and his skin was so dark, there were traces of purple. He did not move and he did not smile. I looked from him to Mom and back again.

"Hello," Mom said.

He dipped his head—once—but otherwise didn't move.

"This is Fizzy Russo," Mom said, placing a hand on my back as if to usher me forward.

The man extended one massive arm, motioning for me to enter the station.

I smiled at him.

Nothing.

"You already know my name. What's yours?" I asked, moving behind the table, which was covered with a clean, white tablecloth.

"Smiley," he said, without a hint of a smile.

I wanted to giggle, but didn't dare. So, instead, I turned away, slipping into the *Southern Living* chef's coat that had been left for me and rolling up the sleeves, because they were too long. Even so, as I smoothed out the fabric and ran my fingers over the sewn-on lettering, I hoped I'd get to keep the coat.

At first, the live audience milling around made me nervous, but then I convinced myself they were no different from my television studio audience, and that helped a lot—because I'd had lots of practice with my studio audience, and they adore me. So after that, it was a snap. My fancy stainless-steel oven was already preheated for me. My ingredients were already prepared and arranged in bowls—yes, the bowls matched and they were clear glass! All I had to do was mix and layer the ingredients and pop the whole thing into the oven. Before I knew it, I was done.

When the judges started coming around and tasting everything, I have to admit I felt a little sick—not homesick but nervous-sick. When they tasted my lasagna, none of them smiled or nodded or gave any indication that they'd just tasted something they liked. They'd take a bite, write something down on their clipboards, and then take another bite. The bald judge even used his fork to peel back each layer and inspect the lasagna, and I just knew he was about to say something like, *What did you put in this . . . to ruin it?* But he didn't say a word. "It's much better the second day," I assured him. The red-haired lady judge gave me a sympathetic little smile. I was worried, and not altogether sure I liked the folks from *Southern Living*.

I liked them even less when a voice came booming through the speakers in the ballroom, thanking everyone for coming and inviting us all to come back at six, to see the winners announced. With a loud, dramatic sigh—which involved my shoulders—I began tidying up my station.

"Don't do that," Smiley said, as if I'd started tossing food onto the floor instead of into the trash can.

I stopped immediately.

Once again, Smiley held out a beefy arm to indicate that I should come out from my station, which I did.

Mom and I went back to our room, where she thought we both "ought to try to rest some." She rested, while I lay on my back, clutching the sheets in two tight fists and staring at the ceiling, barely blinking. I wondered if any of the other cook-off contestants really *needed* to win as badly as I did. Then I tried to imagine what it might feel like if I won, but couldn't. After that, I tried to imagine what it would feel like if I lost: I figured things would be the same as they were before, only with another layer of disappointment—in me. I couldn't stand the thought of things being the same, couldn't stand the thought of being even more disappointing than I already was. *Could. Not. Stand. It.* So I squeezed my eyes shut and prayed, *Help me, help me, help me, pleeeease,* until the phone rang with our wake-up call.

Mom and I stood in the crowded ballroom wearing our best dresses. My heart pounded in my ears. My breathing was too fast. Little dots of color danced in front of my eyes. I thought I might black out. And then I heard my name—I thought. I turned to look at Mom, whose eyes were already on me. She was smiling. "Um, did they just—" I started, but then I heard the announcer say, "Fizzy Russo, where are you?" A big arm cut through the people standing around me, offering itself, and my eyes followed it to the face of its owner: Smiley. I latched on to his arm, and was led through the crowd, up the stairs, and onto the stage. Had I won something? I didn't know. *They're probably just introducing all the contestants,* I told myself.

But then the man in the tuxedo, who was holding the microphone, said, "Congratulations, Fizzy."

I wanted to say, *What for?* but didn't want to sound dumb, so I only smiled.

"Tell us, how old are you?"

"Twelve."

The audience erupted in applause.

I smiled some more. The lights onstage felt warm on my skin, like sunshine.

The announcer said, "Winning the Family Favorites category is a big accomplishment, but it seems like an even bigger accomplishment for a twelve-year-old."

I'd won my category! I immediately began searching the crowd for Mom's face.

"Tell us, how do you feel? What's going through your mind right now?" the announcer said.

"Um . . . well . . . I don't see my mom—could she come up here with me?" I asked.

Everybody laughed.

Smiley brought Mom up onstage, where she wrapped a protective arm around my shoulders, stood very straight, and beamed with pride and joy. Mom told the audience how I loved to cook, how I'd worked really hard preparing my entries for the cook-off, and stuff like that. And I knew she loved me way more than she ever did before. I was glad.

I was also glad to be presented with a cardboard check for $10,000 that was almost as big as I was!

I felt bad for Ms. Marla, though. Until she and her Sweet Potato Cake were announced not only as the Southern Desserts winner, but as the Grand Prize winner! Then Ms. Marla joined us onstage and I felt so happy, I could hardly stand still! Honest. I mean, I'd choose cake over lasagna every day and twice on Sunday—who wouldn't?

Smiley motioned to Mom and me from behind the curtain, off to the side of the stage.

Then, while Ms. Marla was telling the audience how she planned to use part of her winnings to fix a plumbing problem known as "ruts" in her house, Mom and I slipped backstage, where Smiley was waiting.

"What, exactly, is 'ruts'?" the announcer asked.

Ms. Marla said, "You know . . . *tree* ruts—in the pipes."

"Tree roots?" the announcer said.

"That's what I said," Ms. Marla told him.

The audience laughed and laughed, while Smiley escorted us from backstage, down some stairs, into the basement, down a long hallway, to an elevator. As we walked, Smiley said, "People don't mean any harm, but they tend to get a little loud, a little pushy, a little grabby around the winners when it's all over—she's so small—you understand?"

"I do," Mom said. "Thank you."

We walked the rest of the way in silence. I really wanted to ask Smiley why in the world he was called Smiley, but I knew Mom would think that was rude, so I didn't.

When we arrived at the door to our room, Mom and I thanked Smiley.

He dipped his head, once, and then pulled some white fabric out from under his black jacket and handed it to me.

I unfolded it: my *Southern Living* chef's coat! I threw my arms around Smiley and squeezed.

When I pulled back from the hug, he offered me the biggest, whitest, most dazzling smile I'd ever seen—it completely took over his face—he was *all* . . . Smiley. I grinned, thanked him again, and then hurried into our room. After all, Mom and I had a lot of phone calls to make.

Aunt Liz said, once again, that she wasn't surprised.

Suzanne squealed and dropped the phone when I told her, and I could hear her shouting for Dad to pick up right away.

Miyoko said, "You are a black belt in cooking," and we giggled.

Zach said, "It was probably that lucky alarm clock I gave you," and we laughed.

Our last phone call was to Keene and I let Mom make it.

After she told him the news, Mom held the phone out to me and said, "He wants to talk to you, Fizzy."

I took the phone, sat down on the bed, and said, "Hello?"

"Congratulations, Fizzy," Keene said.

"Thank you."

"I have to be honest: I wasn't sure you could do it."

"I know," I said. "That's part of the reason I won."

Silence hummed over the telephone line, until Keene said, "What do you mean?"

I took a deep breath and said, "I knew you didn't think I could do it, so I had to prove you wrong." I looked over at Mom to see if she was listening.

Mom pretended to be fixing her hair in the gold-framed mirror, but I could tell she was listening.

"Well, I've never in my life been happier about being wrong," Keene said.

"You want to know something?" I said, turning away from Mom, in a small attempt at privacy. "I almost went to bed instead of working on that lasagna. I almost didn't make it, didn't send it in. I *wouldn't have*, if you hadn't doubted me."

It was then that I remembered Mrs. Sloan's words: *Sometimes God blesses us with people we never wanted or asked for— because He knows we need them, even if we don't.* "So, um . . . thanks for that, Keene," I said, and I meant it.

Keene was quiet for a few seconds. Then he chuckled and said, "Glad I could help."

When I was off the phone, Mom said, "Was I part, too?"

"Ma'am?" I said.

Mom put a hand on her hip and said, "You said Keene was part of the reason you won. Was I part, too?"

"Yes, ma'am," I said. "I wanted to prove Keene wrong just as much as I wanted to prove you right . . . for believing in me."

Mom smiled. Then she headed for the door as she said, "I'll be right back. I have to go downstairs and pick up the normal-size check so we can take it to the bank."

"We can't take the big one?" I whined, disappointed because I'd really been looking forward to everyone at the bank knowing I'd won at the *Southern Living* Cook-Off.

Mom shook her head and laughed on her way out the door.

Maybe I can wear my chef's coat to the bank, I told myself. *And to the grocery store. And to school!*

I sat on the bed, thinking things over. For the first time, I understood that sometimes, someone doubting you is as helpful as someone believing in you. I didn't know that before.

Then I sat there waiting for the change. I mean, I was *somebody* now. I'd *won*. My dream had come true and it was a whopper of a dream. That kind of thing had to change a person, didn't it? Of course it did!

So I sat there, waiting for greatness to descend on me, waiting to become a bigger, better person.

But I was still just me.

Chapter 38

I couldn't believe it when Mom asked where I wanted to eat dinner to celebrate my win on my last night in the city of Charleston.

"*Here*," I said, as if she should've known.

"Oh no, Fizzy," Mom said. "You don't mean you want to order room service again, do you?"

But I did.

After we'd eaten and I'd had plenty of time to bid farewell to those beautiful silver lids, Mom pushed the room service cart out into the hallway.

I was still just me, I realized again, and I was beginning to feel really sad about it.

When Mom came back into our room, I said, "You love me more, right?"

"More?" she said.

"Yes, you love me more now . . . for winning the cook-off . . . right?"

Mom looked confused, but she answered anyway. "No, Fizzy. I love you just the same. I told you before, there's nothing that could make me love you any less, and maybe I should've told you then, there's nothing that could make me love you any more either."

"So nothing's changed? I'm still just *me*?" I whined.

"What's wrong with being you?" Mom asked.

I gave her a dark look.

Mom stopped moving and waited for an answer.

"You know what's wrong with me," I said accusingly. "*You. Know.*"

Mom came and lowered herself onto the edge of the other bed so that she sat across from me. "No, Fizzy, I don't," she said softly.

"But you have to!" I shrieked. "Because if you don't know what's wrong with me . . . then . . ." *How will I ever get fixed? How will I ever be okay?* That's what I was thinking, only I couldn't say it, because my throat closed; my face crumpled, and tears spouted from my eyes.

Mom rushed to put her arms around me.

"Don't!" I practically shouted, pushing her arms away. "Don't touch me! And don't lie to me!"

"Fizzy, I don't know what you're talking about. There's nothing wrong with you," Mom said, staring at me through wide, unblinking eyes.

I shot up off the bed. "There is!" I wailed. "There *has* to be!"

"But . . . why?"

I threw my arms out at my sides and screamed, "Because even my own family doesn't want me!" I accidently hit the lamp on the nightstand between the two beds. The shade popped off, hit the side of the bed, and bounced onto the floor.

"Calm down, Fizzy. That's not true," Mom said, bending and reaching for the lampshade.

I only meant to kick the lampshade out of her reach, but it went flying off, hit the window, and rolled across the floor. "Yes, it *is* true!" I yelled.

Mom gave up on the lampshade then and, instead, reached up, grabbed both of my hands, and tried to pull me down on the bed beside her. "*No*," she started, but I yanked my hands loose, turned my back on her, threw myself facedown on the other bed, and cried and cried.

Mom got up and moved to the other bed, too, where she placed a gentle hand on my back and let me kick and cry and scream—into a pillow—until I'd pretty much worn myself out.

When I finally began to quiet down, Mom got up, went to the bathroom, and came back carrying the tissue box, which she handed to me.

I pushed myself up into a sitting position and mumbled, "Thank you."

Mom nodded, sat down beside me, and waited.

I wiped my face and blew my nose.

"Now," Mom said. "May I ask why you think we don't want you?"

I shrugged. "I guess I'm not pretty enough or smart enough or good enough—or maybe I just make too many mistakes—that's probably it."

Mom shook her head. "That's not what I was asking, but let

me just say that you *are* pretty enough and smart enough and good enough ... and, Fizzy, *everybody* makes mistakes."

"Yeah, except I can't make them anymore," I said.

"Why not?" Mom asked.

I felt my chin tremble, so I slapped a hand over it and waited. When it stopped, I said, "Because Keene lets me live in your house even though he doesn't want to, just like Suzanne lets me stay at Dad's house when she doesn't want to—nobody really *wants* me, and if I make too many mistakes ..." My chin started quivering again and more tears sprang to my eyes, so I had to stop talking.

Mom took my hand in hers and asked, "If you make too many mistakes, then what?"

"Then Keene and Suzanne will stop letting me stay ... and I'll be *homeless*!"

"Fizzy, that will never happen."

"Okay, maybe I won't be homeless," I sobbed, "but I could end up in foster care—Zach's told me about foster care!"

"No, never," Mom said certainly, squeezing my hand. "Fizzy, it may not feel like we're a family yet—you and me and Keene—but that doesn't change the fact that we *are* a family."

I sniffed.

Mom continued, "Families don't keep score. They accept each other, flaws, mistakes, and all. They love and care for each other, not because they're perfect—nobody's perfect—but just because they're family."

I thought about that while fresh tears slid down my cheeks and dropped onto my shirt.

Mom let go of my hand, plucked two tissues from the box in my lap, and handed them to me. "Is this why you've been cleaning and cooking your guts out?"

I buried my face in the tissues and nodded.

"You don't have to do that anymore. Nobody expects you to be perfect. Do you think Keene's perfect? Let me tell you, he isn't, and neither am I."

"I think Suzanne might be almost perfect," I said. "I mean, *really*."

Mom rolled her eyes. "She isn't—trust me."

I knew better than to argue with Mom about Suzanne, so I just sat there forcing myself to breathe deeply and evenly.

A few minutes passed and then Mom said, "What makes you think Keene doesn't want you? Has he told you that?"

"No," I said, "but—"

Mom interrupted, "No, of course not, and he never will."

I told her about the woman I'd overheard at Keene's family reunion, and how she'd gone on and on about how sweet Keene was to let me live with him, as though he'd made some sort of incredible sacrifice or something. "Am I really that terrible?" I asked. "So terrible that people just can't imagine letting me hang around?"

Mom didn't answer me right away, so I turned to look at her.

Her face was pinchy and angry. "Who said that? At the reunion, who said that?"

"I don't know."

Mom took a deep breath and looked at me. "Well, whoever

she is, she obviously doesn't know you—she obviously doesn't know *anything* about *anything*."

It was true that we didn't know each other and that did help a little.

"Has *Keene* done anything to make you think he doesn't want you?"

I shrugged. "He hardly ever talks to me."

"Have you ever thought that maybe he's just as scared as you are, just as afraid of making a mistake or saying something wrong?"

"No, ma'am."

"Well, he is," Mom said.

Wow. I'd never thought of that, but it made sense, didn't it? I mean, Keene didn't usually say—or do—the right things.

"We're all a little nervous," Mom said. "It's just going to take some time for everybody to get used to living together— even me. It'll get better, you'll see."

"But it won't be the same, will it?"

"The same as what?"

"The same as it was with Dad," I said, and then I told her how I felt like I had lost an important grocery bag, the one with all the important ingredients. *For my life.*

"It can be good, Fizzy, but it can never be the same. Because you haven't been given substitutions for an old recipe; you've been given new ingredients for a whole new recipe, a new life."

"But I didn't want a new life," I told the box of tissues in my lap.

"I know," Mom said, putting her arm around me. "But life

changes whether we want it to or not. No matter what. Life *is* change."

Yuck.

Mom gave me a sympathetic little squeeze. "What about Suzanne?" she asked. "She doesn't talk to you either?"

"Huh? Oh. No, ma'am, she does." I told my mom about the picture in the bedroom then.

Mom thought about it and then finally admitted, "I'm sorry. I don't know the answer to that one. I could understand if you hadn't been there for the pictures, but . . ."

I nodded.

"But I *do* know your father well enough to know he feels the same way I do about family. And if Suzanne didn't feel that way, too, he wouldn't have married her. Trust him, okay?"

I guessed I didn't have any reason not to trust my dad.

"And if the picture bothers you that much, then I think you should talk to your dad about it."

"Maybe."

Mom snapped her fingers. "I know exactly what we need."

"What?"

"Brownie batter."

I couldn't disagree.

But room service disagreed—strongly. When Mom called them, they told her they absolutely could not, under any circumstances, bring us brownie batter, due to the raw eggs in it from which we might contract salmonella poisoning, for which they would be held responsible—in a court of law. Mom told them we'd settle for two warm brownies and some milk.

Meanwhile, I disposed of all my snotty tissues, put the tissue box back in the bathroom, and washed my face with cold water—so the room service person would know that I was fine.

Mom parked the room service cart at the end of a bed and pulled the desk chair up to one side. I sat down on the other side of the cart, on the edge of the bed. Once we were into our brownies, Mom ventured, "Anything else on your mind?"

I shrugged. "I guess not."

"I'm not convinced," Mom said, and she spooned warm brownie into her mouth.

I put my spoon down. "It's just that . . . you always say love means compromising."

Mom swallowed. "That's right."

"But you aren't compromising with *me*, Mom. I used to cook dinner all the time, and now I hardly ever get to. I used to watch TV with you, and now I never do—now you watch with Keene. I used to have lots of time with you, but now I never have any time with you."

Mom stopped eating and lowered her eyes while she thought about it. "You're right," she said sadly. "And I did notice you withdrawing to your room more and more . . . but I thought—hoped—it was just your age."

"Why?"

"I guess . . . I guess I'm scared, too. I just want everything and everybody to be all right. But you aren't." She closed her eyes and shook her head in a way that made her seem . . . *defeated*.

I felt bad then and sort of wished I hadn't said those things.

When Mom looked up at me, her eyes were pleading. "I wish you'd said something sooner, Fizzy. Why didn't you?"

I shrugged. "You're my mom."

"Yes . . . and?"

"And I'm your daughter. I'm not supposed to question you or be disrespectful or rude—it's not like I can just start taking out my earrings, you know?"

"Taking out your earrings?" Mom repeated, looking confused. "What does that mean?"

"Oh, that's how you know when two girls are about to fight at school—one of them will start taking out her earrings."

"At LVMS?"

"No, ma'am, I saw it at my old school. Once."

Mom thought about this, laughed a breathy little laugh, and shook her head. "Fizzy, I want you to keep your earrings on—*always*—and to be respectful, but more than that, I want to know what's going on with you. I'd rather you talk to me than be polite to me, okay?"

"Okay."

"And you're right about compromise: You should be allowed to do all of those things more often. I'll work on it."

"Okay," I said, feeling a lot better all of a sudden—probably it was the chocolate.

"See?" Mom said. "I'm not perfect. I've made some mistakes. Do you want me to move out of the house?"

I smiled. "No, ma'am."

"Are you sure? Are you sure you can forgive me?"

"Yes, ma'am. I forgive you and I love you."

Mom smiled. "Of course you do, because *that's what families do*."

I nodded my understanding.

"Anything else?"

"I still want to take karate lessons," I offered.

"All right, I'll call about karate lessons as soon as we get home. I promise. Is that all?"

I searched my brain and found one last—bothersome—thing: "Miyoko's mom is mean."

"To you?"

"No, mostly to Miyoko."

Mom sighed. "Sweet pea, we all do the best we can for our children—even Miyoko's mom—I'm sure she's doing what she thinks is best for Miyoko."

"But it isn't . . . best."

"If you're sure about that, then remember her mistakes and try not to make the same ones when you're a mother."

"Okay," I said, "but I wish I could help Miyoko."

"You can, you *do*, just by being her friend and loving her."

That didn't seem like enough.

"Just remember, Fizzy, we all make mistakes, even when we're trying our hardest and doing our very best."

"Yes, ma'am."

After we'd finished our brownies and milk, Mom said, "Now go brush your teeth and get ready for bed."

I knew for sure then that I was still me, so I said, "I can't believe the cook-off is over and I'm still just plain old freckle-faced Fizzy Russo, the leftover kid . . . who is nobody special at all!"

Mom stopped pushing the room service cart and stood up straight. "Did you say 'leftover'?"

I hadn't realized I'd said it, but I had. I nodded. Then I explained the thing about leftover spaghetti and leftover kids.

Tears filled Mom's eyes as she stood frozen and she was quiet for a good long while.

Then she said, "I guess you are a kind of leftover, Fizzy, but you're certainly not spaghetti. If you're a leftover, then you're lasagna. You get better every day. You learn every day—sometimes by making mistakes—and you get better."

It's true that lasagna gets better every day that it sits in the fridge. I have no idea how or why this works, but it does. Even so, I had never considered lasagna leftovers; I thought of "refrigerate for twenty-four hours" more like the last step in the process of making really excellent lasagna. But I guessed that twenty-four hours in the refrigerator technically made anything leftovers—even lasagna. Who knew?

Mom continued, "Think about it, Fizzy. Just a year ago, you were merely experimenting with our dinners, and today you are a *Southern Living* Cook-Off winner."

I nodded.

Mom sat down beside me on the edge of the bed and covered my hand with hers. "But none of that has anything to do with how much I love you or why," she said. "Fizzy, I love you just because you're mine. You don't have to do anything to earn it. It just is . . . it always will be. Okay?"

Relieved, I exhaled and said, "Okay."

Mom smiled, patted my hand, and stood. "Go get ready for bed now."

I thought about what Mom had said and couldn't help wondering if the reason Keene *didn't* love me was just because I *wasn't* his—because that wouldn't be my fault—right? That's what I was thinking as I rooted around in my bathroom bag, looking for toothpaste. When I came across a big, black plastic spider, I plucked it out and held it. Did Suzanne love me a little bit? Even though I wasn't hers? I thought she might, and not in a way that was forced—by Dad. I thought Suzanne might actually love me, all on her own, just because she found me . . . lovable. It *was* hard to believe, but . . . if true, then . . .

Maybe, in time, Keene would love me, too. And if he didn't, well, that probably wouldn't be my fault—and that's what's important.

Chapter 39

On the last Sunday of summer vacation, I was at Dad's, holding Baby Robert and humming his favorite lullaby, "Nen-neko yo."

Now that Baby Robert was past all that awful colic, he was happy. He *always* smiled with pure delight when he saw me—like when Aunt Liz sees me. With Baby Robert, there was no question that I belonged; he made me feel like part of the family—an *important* part of the family. I loved him for that, and for his cheerful chirpy sounds, his warm, sweet baby smell, his fat pink cheeks. I just loved him.

Dad took Baby Robert from my arms and I let him—even though I didn't really want to—while Suzanne continued to clear our dishes from the table. Then he sent me upstairs to collect my suitcase and stuff.

I did as I was told, but on my way back down, I stopped in Dad and Suzanne's bedroom. As I stood there looking at the picture of Dad, Suzanne, and Baby Robert, I decided once and for all not to ask Dad about it. I'd given it a lot of thought: What did I hope to get out of it? What could Dad do or say to make me feel better? Nothing. I mean, if having a picture of the three of them—the three people who lived in their house and were part of their family every day—made Dad and Suzanne happy, then I wanted them to have the picture. I wanted them to be happy, because I love them. Both of them.

But even if I hadn't cared about their happiness, making them take down the picture wouldn't change the reason they put it up in the first place. It wouldn't change who I am or how they feel about me or the way I feel about them any more than winning a cooking contest had—which is not at all. I mean, if Dad and Suzanne love Baby Robert way more than they love me, there isn't much I can do about that, is there? Besides, I could understand: Baby Robert really is the best thing since cupcakes.

I guessed I could've asked Dad about this, but I really didn't want to know *for sure* that he loves Baby Robert more than me. And if he said he didn't, would I believe him? I mean, what kind of parent tells you he loves your brother way more than he loves you?

That's what I was thinking when Dad snuck up behind me and cleared his throat to announce himself.

I lowered my head, feeling somehow ashamed that I'd been caught looking at the picture.

Dad placed a warm hand on each of my shoulders and said, "Sometimes, when you love somebody, and you know they love you, you let little things pass."

I nodded at the rug.

"Suzanne said she didn't see the point of buying and hanging two identical photos any more than she would see the point of buying and hanging two identical paintings in the house."

I thought about this.

Again, Dad said, "Sometimes, when you love somebody and you know they love you, you let little things pass. You know we love you, right?"

I decided to let it pass. I wasn't going to think about it anymore. I really wasn't. Because family doesn't keep score, which is why I also tore all Suzanne- and Keene-related lists out of my journal and threw them away—because they're family.

On the way home, Dad stopped and took me shopping for school supplies. Now, there is a big difference between shopping with my mom and shopping with my dad. Mom is practical. She wants things that are well made and built to last. But Dad just wants to be done. So he didn't inspect the things I picked out. He barely even looked at them. Instead, he just kept saying, "Put it in the cart." And that is how I got the coolest school supplies ever!

Keene was out of town on business, but I told him about my supplies when he called, and he sounded happy for me. Then he asked where my shoes were.

I panicked as I tried to think: *Were they by the front door? Again?*

"Fizzy—" Keene started, but I interrupted him.

"Keep your earrings on," I said. "I'll put them away as soon as I hang up."

Keene chuckled. Mom and I say "keep your earrings on" all the time now—instead of "keep your panties on" or "calm down"—and Keene is in on the joke.

I put all of my shoes away as soon as we hung up.

After that, Zach called to say he was having a campfire in his backyard tonight and there would be roasted hot dogs and s'mores. He wanted to know if I could come.

"Is Miyoko coming?" I asked.

"Yeah."

"Anybody else?"

"Like who?"

"Like . . . I don't know . . . Buffy Lawson?" I teased.

"Fizzy, I gave you an *alarm clock*," Zach said, as if "alarm clock" and "engagement ring" meant exactly the same thing.

I smiled and said, "I bet you give alarm clocks to all the girls."

"Just you," Zach said.

I agreed to come, so long as Zach agreed not to build the campfire without me. I've learned lots of fire-building techniques from *Survivor Steve*, and I figured this was my big chance to try some of them out.

Keene, Mom, and I watch *Survivor Steve* together now. According to Survivor Steve, I'm a survivor, too, because I accept change, adapt to it, and move forward—quickly—and I'm getting better at it.

Even Mom seems to recognize this. She let me buy my own phone with some of my cook-off winnings because she said that I'd already learned to accept that I can't always get what I want "in other ways." I also bought two pairs of designer jeans and some new flannel shirts. But Mom wouldn't let me buy any—mini-miracle—makeup. She said that I'm still too young for makeup and that I don't need it. So, the rest of the money went into a college fund for me—which, of course, is really a culinary school fund—but I've accepted, adapted, and moved forward. I now think of *college* as the code word for "culinary school."

So, for the most part, things are pretty good. I cook dinner twice a week, and Mom and I do the grocery shopping—just the two of us—every other Saturday morning. Granted, though, the ugly, puke-green recliner is still in our living room and I still hate it. But maybe it'll grow on me—or match something someday. Things change.

Even I change. I actually *like* leftovers now. I've liked them ever since I discovered this website where I can type in all the ingredients I have—leftovers—and then it spits out all the new meals I can make with them. Some of the best, most beautiful meals I've created lately have been leftovers! Yesterday, for example, I chopped up our leftover chicken and made the best chicken salad in chicken-salad history, with diced pickles and grapes, which are the perfect balance of sweet and sour—who would've guessed?—and toasted slivered almonds to add a crunchy texture! And I love brussels sprouts—when they're tossed in olive oil, garlic, and spicy mustard, roasted until crisp, and then salted—they're best salty and hot, like French fries.

When I'm happy, I try to really pay attention, because things are bound to change. When I'm unhappy, I try to wait it out, because things are bound to change.

But at all times, I try to keep my list of Notes to Self in mind:

1) *Church shoes are important to Dad, too—remember them—nobody likes Sunday sneakers!*
2) *Chewing gum is not only unattractive, it's dangerous!*

3) When you're scared, don't talk—you'll probably say something obnoxious.
4) It's never smart to mess with a girl who has a pimple with an eyeball—or her friends.
5) Everybody else is too worried about their own Ogles to notice yours. Probably.
6) Suitcases don't have to say, "My family is a big, broken mess and so am I!" They can also say, "I am a totally normal person who has friends, and I'm sleeping over with one of them! Yay!"
7) Makeup is a mini-miracle. Get some. (As soon as you turn sixteen, Mom says.)
8) If you're ever standing in your front yard—or anywhere, really—wearing only your underpants, DO NOT SCREAM! Because anybody who isn't looking will start.
9) More cakes and less mistakes. (Just do your best, Mom says.)
10) Punishing others by not allowing them to help you isn't a good punishment—for them—it's a great punishment for YOU.
11) Fix the problem, not the blame. (I learned this Japanese proverb from my karate teacher, who says, "Figuring out who to blame doesn't solve the problem; figure out how to solve the problem instead." He is very smart.)

Acknowledgments

To me, this book is (more) proof that if you give God your ashes, He'll grow something beautiful in them; my heart overflows with joy and gratitude at all He has done in me, through me, and for me.

I remain grateful for and to my amazing family, friends, and colleagues. Words cannot adequately express my love and appreciation to my husband, Mark; my daughter, Laurel Grace; my niece, Airen; my sister, Sarah; and my parents (all of them), who generously read and re-read for me, and to Dr. Susan Couzens, cold-reader, soothsayer, pray-er, and friend extraordinaire. In addition to reading for me repeatedly, all of these people believed in me so strongly that when my own belief was lagging, theirs carried me forward.

Likewise, my agent, Emily van Beek, also believed, and her confidence in me literally changed my life. I'm pretty sure that without Emily, all my novels would be confined to shoe boxes under my bed . . . well, except for the rare occasion when I might get one out to try trading it for something slightly more useful—like a bucket of chicken. Thank you, Emily, for not only finding a house for my books, but for finding a house that feels like home.

It was in Nancy Paulsen's house that *The Thing About Leftovers* became the book it is today. Nancy knows when to ask questions, when to give answers, when to push forward, when to pull back, and when to let go—and she does it all while being encouraging! She also knows when to laugh—the importance of this quality cannot be overstated. Somehow, Nancy made me love writing even more—I didn't know that was possible! I will always be grateful to Nancy and to everyone in her house who lent their time, talent, and skills to this book, including Sara LaFleur, Cindy Howle, Anne Heausler, Jeanine Henderson Murch, Irene Vandervoort, and Annie Ericsson.

A big thank-you to R. David Clark, who is always ready and willing to give me information about the legal system.

Thanks to my friend Jennifer Releford, who coined the phrase "Big Booty Judy," who always makes me laugh, and who keeps me looking (relatively) presentable—by doing my hair.

Thanks to Alaine Carpenter, Jennifer Owen, and Katrina Williams, three of the best friends a book—and an author—could ever have.

Thanks to Fizzy Ramsey, P.A., who kindly agreed to share her first name with my main character.

Last but certainly not least, I thank YOU, my friend and reader, for taking this journey with Fizzy and me. We couldn't do it without you. Really, we couldn't. At all. So thank you, thank you, thank you!